W9-ANY-172

Classification of Remotely Sensed Images

The purpose of computation is insight not numbers

Hamming

Classification of Remotely Sensed Images

Ian L Thomas
Department of Scientific and Industrial Research, Lower Hutt, New Zealand

Vivien M Benning
Department of Lands and Survey, Wellington, New Zealand

Neville P Ching
New Zealand Forest Service, Wellington, New Zealand

Adam Hilger, Bristol

Soc
G
70.4
T46
1987

British Library Cataloguing in Publication Data

Thomas, Ian L.
   Classification of remotely sensed images.
   1. Remote sensing–Data processing
   2. Image processing–Digital techniques
   I. Title   II. Benning, Vivien M.
   III. Ching, Neville P.
   621.36′78′0285   G70.4

   ISBN 0-85274-496-X

Consultant Editor: **Professor M H Rogers**, University of Bristol

Published under the Adam Hilger Imprint by IOP Publishing Limited
Techno House, Redcliffe Way, Bristol BS1 6NX, England

Typeset by KEYTEC, Bridport, Dorset, England
Printed in Great Britain by J W Arrowsmith Ltd, Bristol

Contents

Preface

Most users of remote sensing data are now familiar with relating grey tones or colours from photographs to what actually exists on the ground. The emphasis however is now changing and extending more into handling data on interactive processing systems and implementing the various classification techniques. Such classifications take the multiband data and, usually with user guidance, produce a coded thematic map for review by the user.

It is our purpose here to assist discipline-oriented users in the use of such interactive classification systems. We contend that such discipline-oriented users should not be turned into computer operators. They should bring their particular skills to a 'user-friendly' system and make the best use of it. This book is directed at helping such users approach the machinery with more confidence and to act as a ready reference whilst they are using the systems.

This book has been written with the needs of farmers, cartographers, foresters, hydrologists, geographers, etc, firmly in mind. The project was undertaken with a view to making it possible for their use of computer-based interpretation systems of remotely sensed data easier, more informed and more integrated into their needs and existing discipline skills.

The classification technique we ultimately focus on is one in common use, being the maximum likelihood classification system. However the general principles outlined here in the classification process apply to most techniques. The concepts are illustrated by projects conducted in New Zealand, although the concepts are universal in nature and are considered to be applicable to any part of the world.

Two databases are considered. The first is from Landsat and the second is provided by an aircraft scanner system. Data such as are considered from the Landsat 2 system are available from most parts of the world in archival form. This has been made possible by the tape recorders aboard the older series of the Landsat system. With such ready data availablility most users should be able to start out with data such as described here. These digital foundations, when laid, can be built upon with the analyses based on later Landsat and other data streams. The use of aricraft scanners is becoming more frequent for localised surveys. It is noted that much the same analysis techniques apply to these systems as apply to the earlier Landsats considered initially. It is hoped that users following this analysis review and these example studies can approach all the digital data streams presently before us with confidence.

Whilst the IBM ERMAN package was the one used in these examples, the generalised techniques are common to all analysis systems. All the techniques considered here are appropriate to a wide variety of present and upcoming analysis facilities.

This introduction to the classification of multichannel remotely sensed data is based upon a report entitled 'Computer Classification of Landsat and Aircraft Scanner Images – The Collected Papers of the 'ERMAN Project', edited by I L Thomas and published as *Report No 766* by the Physics and Engineering Laboratory, Department of Scientific and Industrial Research, New Zealand, in October 1982. That report covered Joint Research Program Agreements with IBM New Zealand Ltd and included a variety of aspects. The material in this book has been condensed to address the aspects felt useful to those who are starting in the multichannel classification of remotely sensed data. As such, each of the chapters here is based upon chapters of the Project Report (see the Appendix, 'Sources of material').

It is our pleasure to acknowledge the author teams of the Project Report. The overall responsibility for the direction, content and all conclusions drawn here, however, firmly resides in our hands.

Ian L Thomas
Vivien M Benning
Neville P Ching
Wellington, NZ
June 1986

Acknowledgments

The authors wish to acknowledge the support of the following. The New Zealand Government, especially the Department of Scientific and Industrial Research, New Zealand Forest Service and the Department of Lands and Survey, for supporting this work; IBM New Zealand Ltd and IBM Australia Ltd for making the ERMAN system and technical support available; Mrs C Keppel and Mrs S Coburn, DSIR, New Zealand for typing manuscripts and Dr J Buckingham and Miss A Southern for acting as referees.

Affiliations

The original authors had the following affiliations at the time of the ERMAN project:

I L Thomas, G McK Allcock, A F Cresswell, P J Ellis, J E Lukens†, M J McDonnell, S M Timmins

Physics and Engineering Lab., Department of Scientific and Industrial Research, Private Bag, Lower Hutt, New Zealand
†Fulbright Scholar from Rhode Island School of Design, Providence, Rhode Island, USA

V M Benning, N M Davis

Dept of Lands and Survey, Private Bag, Wellington, New Zealand

N P Ching

New Zealand Forest Service, Private Bag, Wellington, New Zealand

D W Beach, B J Winters

IBM Australia Ltd, Kent St. and Bradfield Highway, Sydney, New South Wales, Australia

R L Bennetts

Ministry of Agriculture and Fisheries, Darfield, Canterbury, New Zealand

1 Photointerpretation to Digital Classification

1.1 The Purpose of this Book

This book is directed to the user who wishes to start applying digital remote sensing technology to the mapping of ground cover classes. Such a user may be a cartographer, a forester, a farmer, an agricultural adviser, a hydrologist, a geologist, a glaciologist, etc, as well as those computer-oriented remote sensing technologists who wish to work with other discipline-oriented users whilst starting to apply the technology to the problem.

We want to take a user from a background of essentially photointerpreting remotely sensed imagery (such as that obtained from Landsat, or an aircraft scanner etc) through to confidently loading tapes and executing an informed digital classification and comparing the results. Only through this latter comparison will the user become more proficient in interpretation; more use can then be made of the data and better management of ground cover resources will eventuate. We do not look to an approach that simply rests on an optimistic pressing of buttons on an interactive computing system. This can so easily generate an amazing and numbing kaleidoscope of colours. Rather, we want the user to embark upon a more methodical and reasoned approach to ordering his/her needs, evaluating the capability of the data to satisfy those needs and then entering the classification process with an understanding of each step along the way. Then, following the completion of the classification, the worth of the product can and must be evaluated.

We assume that the user starting along this digital 'yellow brick road' towards the computerised equivalent of the Wizard of Oz is familar with relating remote sensing photographic products to what is actually on the ground in terms of vegetation stands and textural/structural detail. This qualitative interpretative background is here extended into the digital area with a more quantitative product being the result.

This book is written as a sequential progression from suggesting methods that could help the user order his/her analysis needs and comprehend how a spectral signature is produced, through the necessary theory, to examples

1

of studies that have actually been performed using these approaches. This book can also be used as a reference for questions that arise when using, or planning the use of, an interactive analysis system. The detailed cross-referencing in the index can hopefully support this rapid location of a topic and aid the inquiring user. Similarly, the inclusion of a digitally-oriented glossary will, we believe, aid users to locate quickly the appropriate terminology. To further aid this ready-reference type of access by the user, the structure of the book is divided into topic areas. Most users of remote sensing are now familiar with the basics of the Landsat system so no more is included than a very condensed restatement of the salient points. Conversely, few users are aware of some of the problems involved with acquiring and correcting aircraft scanner data. This consequently is treated in greater detail. Also, most users have two thoughts before them as they approach an analysis task. The first is the project objective (agriculture, forestry, etc) and the second is an awareness of the data types that will be used. These data streams may range over Landsat to map systems, or even into a fully integrated Geographic Information System approach. However, at the moment most users seem to approach the task with a focus on either Landsat, spot or an aircraft scanner, and few come to it with an integrated view. We feel this latter approach will come but that it must be based on sound understanding of the handling of individual data streams. Consequently we localise our treatments to individual data streams and topic areas. Accordingly, chapters within the book are of varying length as indicated by our view of the topics and our perception of the usual users' background. We also felt that more specific chapters would assist users in more quickly locating appropriate material as they sit in front of the display screens.

We have tried to present the necessary detail to support a user's probe of the theory in such a way that they can understand and follow the steps as they approach this field of computerised analysis. We aim to help users question the theory behind what they are doing on the system. From this, we contend, will come a more informed user and thus a better product. We do not pretend to present all the theoretical ramifications of the classification task. Our aim is to provide a solid base from which the user can grow and more deeply investigate both the data and the corresponding analysis systems.

As a database we have used four-channel digital data from the Landsat 1, 2, 3 series of spacecraft together with eleven-channel digital data from an aircraft multispectral scanner. We note that such four-channel Landsat data are available of most of the world in archival form since the earlier Landsats carried tape recorders. Consequently it should be possible for any user to have access to the type of data that are considered here. Again, from this approach, we believe that the user's confidence and capability can grow and the use of the data from later Landsats and other systems will follow. All the outlined techniques apply equally well to these data sources and other systems such as the Landsat 4, 5 . . . (Multispectral Scanner (MSS) and Thematic Mapper) series, the French *Système Probatoire d'Observation de la Terre* (SPOT) system, the NOAA Advanced Very High Resolution Radiometer (AVHRR) data, plus all the similar acquisition systems. Once the data are in the system as a series of counts along a scan line, all the techniques may be similarly applied. In the same way digitised map data (be they topographic, soil, cadastral boundary, meteorological data, or any other) may be similarly combined with the above scan line data and be validly used as another input to the classification process.

We limit ourselves in this work to the supervised classification process, where the user interacts with the data in selecting training fields, evaluating results and setting thresholds etc (see Swain and Davis 1978). It has been our judgement and experience that, as a starting point, the supervised approach can lead to a better product with fewer system iterations. It also helps to familiarise the user more quickly with the data and further aids that user in appreciating the limitations imposed by the data and the system upon their expectations. A better product usually results. Once understanding of the supervised classification approach is achieved it is our contention that the user is then more adequately equipped to drive unsupervised analyses, such as those based on clustering algorithms.

Somewhat for the same reason we concentrate on the parametric classification approach where the statistical distribution for each class is typified by a very limited set of numbers. This is preferred over the non-parametric technique where each data point contributes to a distribution in multichannel space. Once the user appreciates the limitations of the parametric technique then the determination of the bounding surface in multichannel space may be more knowledgeably accomplished and the non-parametric techniques of classification followed with more insight.

The classification technique we dwell on is that of the Maximum Likelihood approach. No real attempt is made to examine in detail the application of other techniques. The Maximum Likelihood is becoming more commonly implemented on systems and the statistics that support it will support the growing understanding of the class data distributions for the user.

1.2 Outline of the Book

The user commonly starts with the statement: 'I want to map the occurrence of this vegetation type over that area'. The question of separating that vegetation type from its neighbours is implicit but the practicalities of such separation have often not been considered.

The sensor system looking down at a region, illuminated usually by sunlight (rather than radar illumination or thermal emission) quantifies the amount of energy passing into its detectors, over a limited spectral range in each channel. Sometimes these spectral ranges may not be the most appropriate to the discrimination, or even mapping, sought by the user.

We commence in trying to bring together these concepts that are so vital to a successful classification. These are the factors in the ground cover that influence the composition of a spectral signature, and the spectral/spatial sampling of the sensor systems. The user too must order his/her classification desires in moving from trying to separate land from sea to attempting to discriminate far finer details within classes. This objective, in seeking finer detail in a class, could be at the level of attempting to discriminate a variety of autumn-sown wheat, sown on nitrogen-enriched soil with adequate moisture, from the same variety, for example, but on soil with 5% less moisture. Some form of hierarchical ordering of these overall user requirements aids the rationalising of these desires with:

(i)   the capabilities of the data stream (spectral and spatial);
(ii) the time of data acquisition;
(iii) the acceptable/unacceptable levels of confusion between related classes that could be tolerated in a final classification.

Any such classification obviously builds upon an initial appreciation of an appropriately enhanced photographic product generated from the digital data. A summary of the various major techniques for producing such grey scale/colour/textural enhancements is presented. From these an agriculturalist, for example, could prepare an enhanced image to focus on farming problems and a forester, a differently enhanced image from the same digital data, to focus on the crop health of his trees. Once the data are so evaluated a more informed planning of the ultimate classification process can result, perhaps aided by preliminary classifications using the less involved and less costly parallelepiped classifier.

Once the ground work has been laid, the way is open to commence the maximum likelihood classification work, perhaps with a selection of synthetic channels which have been derived from the basic channels that are recorded by the sensor system. The major synthetic channels (Band Ratios and Principal Components) are outlined, as is a series of techniques for determining the most effective selection of the channels, within this overall channel set, which produce the required class separability in the maximum likelihood classification. These indices of separability build on the same theory that supports the Maximum Likelihood classifier. They can result in a lesser use of computer resources to achieve equivalent class separation. Their use can also answer the often asked question of 'what is the best single channel to discriminate my classes of interest in a black and white print?'. Similarly the best three channels for a colour composite portrayal may also be deduced in a like manner.

A great deal of present work is done by users on single images, i.e. not combining time sequential imagery. As such it is acceptable as well as less costly in terms of computer resources to rectify and register the single channel classification product rather than the multichannel input dataset. (Obviously if time sequential imagery is employed then such rectification/ registration must precede classification.) Following the discussion of classification, we consider the topic of rectification/registration, which enables users to compare their classified product with their base maps more adequately—either through transparent overlays or through field measurements.

The assessment of any classification product is not an 'add-on' but a vital step in the classification process. A procedure for arriving at this assessment is outlined, as is the overall pathway that leads from the initial user request to a checked and map-corrected classification product.

Following this outline of the supportive theory, four demonstration projects are presented. Each concentrates on different aspects of the application of the technology. Three use Landsat data for general land cover, forestry, and specific agricultural targets, and one uses the aircraft scanner data for a more detailed agricultural study.

1.3   References

Swain P H and Davis S M 1978 (ed.) *Remote Sensing: The Quantitative Approach* (New York: McGraw-Hill)

2   Spectral Signatures and the Sensed Pixel

2.1 Spectral Signatures Introduced

Mason (1981) described spectral signatures as the 'fingerprints of life'. Each type of surface on the earth (e.g. sea, rocks, sand, pasture, forest, cities, snow) reflects sunlight modified by the characteristics of that ground cover. A detection system, such as in a spacecraft, measures and records the amount of energy incident in each of the wavelength, or spectral, bands for each picture element (pixel).

The spectral signature for a ground cover class depends upon the interaction of sunlight with that type of ground cover. The factors in the vegetation that lead to the spectral signature are considered here.

The size of a sensed pixel on the ground will vary between detection systems (spacecraft, aircraft scanner, ground radiometer, human eye etc). As a result the amount of spatial and spectral detail that must be averaged to produce the single number in each spectral band does vary between detection systems. This must be borne in mind when comparing data products from the various sources. The degree of microscopic heterogeneity that must be averaged to produce a macroscopically homogeneous ground cover class for the sensed pixel is a question that must always be before us.

2.2 Spectral Signature of Vegetation

The spectral reflectance of most chlorophyll-containing surfaces is similar (Gates 1970). Thus, a generalised spectral reflectance curve which is distinctive may be derived for green vegetation. Figure 2.1 shows the significant spectral response characteristics of green vegetation. The wavelength regions within which the Landsat system operates are indicated.

It can be seen that the spectral reflectance response from green vegetation is quite variable with wavelength. In the visible blue and red wavelengths, absorption is high with slightly reduced absorption in the green. Very high reflectance and transmittance occurs in the near infrared (0.7 to 1.5 µm) followed by increasing absorption into the far infrared (figure 2.1).

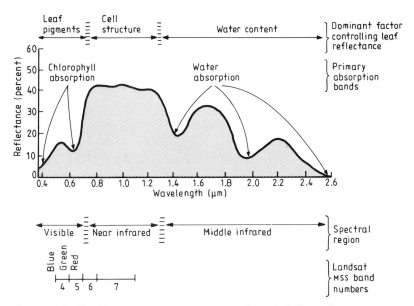

Figure 2.1 Significant spectral response characteristics of green vegetation (after Hoffer 1978).

The low reflectance throughout the visible region is due to absorption of light by leaf pigments. The two troughs in reflectance in the blue and the red correspond to the two chlorophyll absorption bands centred around 0.445 and 0.645 µm (Gates *et al* 1965). Carotenes and xanthophylls are also frequently present in green leaves. They too have an absorption band in the blue portion of the spectrum centred approximately on 0.45 µm (Hoffer 1978).

There is relatively little absorption in the green wavelengths, but reflectance is still comparatively low because of the influence of leaf pigments. This is interesting to note as the human eye can discriminate many different shades of 'green', because of the greater sensitivity of the eye to green wavelengths (Gates 1970).

Variation in the response of green vegetation within the visible wavelengths occurs between different vegetation covers and also over time, even though a basic pattern of reflectance can be detected (figure 2.1). Stress, disease or senescence will cause a decrease in chlorophyll production which will lead to an increase in reflectance in the chlorophyll absorption bands. Decrease in chlorophyll production prior to leaf fall is accompanied by production of large quantities of the red pigment anthocyanin in some species (Hoffer 1978).

In the near infrared green leaves have a high reflectance and transmission. Relatively little energy between 0.7 and 1.5 μm is absorbed (Gates 1970; figure 2.1). There is a greater variation in response recorded from green vegetation in the near infrared than in the visible wavelengths. Plant species or types with distinctive responses in the near infrared wavelengths often have negligibly different reflectance in the visible wavelengths (Hoffer 1978).

In the near infrared, in contrast to the visible wavelengths, the interaction of the incident radiation with the structure of the leaves, and not the pigmentation, largely controls the reflectance (Hoffer 1978). Mature plants, or those with thick leaves or a dense covering of hairs, or a thick waxy cuticle, will have a higher reflectance in the infrared than thin leafed or immature plants.

Infrared energy may be absorbed by the water films associated with the mesophyll cell walls since there is a liquid absorption band centred around 0.98 μm. Thus, as the moisture content of the vegetation decreases, the reflectance over all wavelengths increases (Myers 1975).

The factors contributing to the reflectance characteristics of an individual leaf may also bring about differences in spectral responses between plant types or life cycle stages. For instance, higher infrared reflectance may be recorded from a stand of glossy leafed plants than from a stand of needle leafed dry-loving plants. A thicker cuticle on the leaves of one stand of plants may increase the stand's reflectance relative to another. However, if the leaves also have a high water content, there may be a net depression in recorded reflectance (Hoffer 1978). The relationships between plant condition and stand reflectance are by no means simple.

Colwell (1974) and Hoffer (1978) list several other factors which influence the reflectance of a vegetation canopy: degree of cover, geometry and configuration of the ground cover, characteristics of non-leaf components of the vegetation canopy (stalk, trunks, limbs), characteristics of the background, shadow and any environmental variables. Some of these are discussed below.

Increases in vegetation cover and Leaf Area Index (LAI, the ratio of leaf area to ground area) usually cause an increase in the infrared reflectance of the vegetation canopy and sometimes a decrease in the red reflectance. Reflectance in red wavelengths is dependent upon the percentage cover, the look angle and the solar zenith angle (Colwell 1974, Myers 1970). Near infrared reflectance of the canopy also has a positive correlation with crown closure, and visible reflectance a negative correlation (Colwell 1974). The enhancement of the near infrared reflectance may be attributed in part to the increase in leaf layers with greater percentage cover and LAI (Hoffer 1978). Energy transmitted through the first stratum of leaves and reflected back from a second is partially transmitted through the first layer, increasing reflectance by up to 85%. It must be noted however that as the LAI rises, a decreasing proportion of the total radiation will be reflected from the lowermost leaf strata (Myers 1975).

Suits (1972) predicted that when the leaves of a canopy change from a predominantly horizontal to a predominantly vertical orientation, the visible reflectance of the canopy may increase and the near infrared decrease. This change could be the result of wind, or wilting caused by water stress. In forested areas, stand density and size of trees can cause significant differences in the canopy spectral response (Hoffer 1978).

The reflectance from a vegetation stand is not only an integration of the individual responses of the plants, soil and moisture present in the stand; it is also modified by the quality of the incident light. The spectral distribution of direct sunlight on a clear day has a broad peak of intensity between 0.5 and 1.0 μm. The distribution of cloudlight on an overcast day is quite different, being much lower in intensity and peaking in the visible wavelengths (Gates 1970). The quality of light reaching the lower layers of a stand will be further modified and tend to be richer in infrared radiation than that incident on the top canopy layer.

The relations between the Landsat or aircraft spectral data and vegetation data are complex, especially given that the reflectances recorded by the sensors are an integration of all factors influencing reflectance covering a sizable area on the ground (Chapters 4 and 5). In some vegetation stands, non-green vegetation components or soil may predominate and the recorded reflectance may differ significantly from that expected of green vegetation. However, in general, relatively low reflectance values are expected from green vegetation in the visible wavelength bands recorded by the Landsat Multispectral Scanner (MSS) sensor (MSS bands 4 and 5), and much higher values in the infrared bands (MSS bands 6 and 7). Variations occur within these basic trends due to differences in the geometry, morphology and physiology of leaf, plant and canopy. The season, soil and climatic conditions in which the plants are growing also have an effect.

2.3 References

Colwell J E 1974 Vegetation canopy reflectance *Remote Sensing Environ.* **3** 175
Gates D M, Keegan H J, Schleter J C and Weidner V R 1965 Spectral properties of plants *Appl. Opt.* **4** 11
Gates D M 1970 Physical and physiological properties of plants *Remote Sensing with Special Reference to Agriculture and Forestry* ed. J R Shay (Washington, DC: National Academy of Sciences) ch 5

Hoffer R M 1978 Biological and physical considerations in applying computer-aided analysis techniques to remote sensor data *Remote Sensing: The Quantitative Approach* ed. P H Swain and S M Davis (New York: McGraw-Hill) ch 5

Mason R S 1981 Informal Communicaton, Physics and Engineering Laboratory, New Zealand Dept of Scientific and Industrial Research

Myers V I 1970 Soil, water and plant relations *Remote Sensing with Special Reference to Agriculture and Forestry* ed. J R Shay (Washington, DC: National Academy of Sciences) ch 6

—— 1975 Crops and soils *Manual of Remote Sensing* ed. R G Reeves (Falls Church, VA: American Society of Photogrammetry) ch 22

Suits G H 1972 The calculation of the directional reflectance of a vegetative canopy *Remote Sensing Environ.* **2** 117

3 | Levels of Refinement in a Classification

3.1 Aggregation of Ground Cover Classes into Affinity Groups

Ground cover classes may obviously be aggregated into groups of similar classes: rock, water, forest, agricultural crops. Similarly, these major divisions may be further subdivided. For example, the agricultural crop class may be divided into the subclasses of pasture and cereal crops. Within a major crop type an even finer level of discrimination may be attainable at specific times in the growth cycle. As the time of harvest approaches cereals may be divisible into wheats and barleys.

Finer and finer levels of discrimination may emerge with improved spectral and spatial resolution of the detection system. Similarly, a close awareness of the vegetation characteristics of various crops could permit finer discrimination through data acquisition at opportune points in the season.

A recognition of this hierarchical structure can greatly assist the efficient and effective operation of a classification programme.

The analyst should always recognise the most appropriate level of class subdivision for his/her study. The possibility of confusion between sub-classes at the more detailed levels must also be recognised.

If an appropriate level for the overall classification has *not* been recognised, and with it the desired levels appropriate to other class discriminations, then wasteful energy from the analyst and unproductive use of the computer may result. It is important to assess the appropriate hierarchical levels before commencing ground truth collection for the project.

The purpose of this chapter is to introduce the hierarchical classification structure of Anderson *et al* (1972). This is illustrated by applying it to the New Zealand agricultural study (Chapter 14).

3.2 Level of Classification Refinement

Anderson *et al* (1972) put forward the concept of a multilevel system for classifying land cover. They suggested that land use studies may require

Table 3.1 The Anderson *et al* (1972) two-level classification scheme for land cover. It is primarily directed at interpreting satellite imagery from photographic products. Level I is suggested for immediate recognition from the photographic product alone whilst Level II relies on the addition of either imagery at larger scales or subsidiary data, e.g. topographic map data.

Level I	Level II
01 Urban and built-up land	01 Residential
	02 Commercial and services
	03 Industrial
	04 Extractive
	05 Transportation, communications and utilities
	06 Institutional
	07 Strip and clustered settlement
	08 Mixed
	09 Open and other
02 Agricultural land	01 Cropland and pasture
	02 Orchards, groves, bush fruits, vineyards and horticultural areas
	03 Feeding operations
	04 Other
03 Rangeland	01 Grass
	02 Savannas (Palmetto prairies)
	03 Chaparral
	04 Desert shrub
04 Forest land	01 Deciduous
	02 Evergreen (coniferous and other)
	03 Mixed
05 Water	01 Streams and waterways
	02 Lakes
	03 Reservoirs
	04 Bays and estuaries
	05 Other
06 Non-forested wetland	01 Vegetated
	02 Bare
07 Barren land	01 Salt flats
	02 Beaches
	03 Sand other than beaches
	04 Bare exposed rock
	05 Other
08 Tundra	01 Tundra
09 Permanent snow and icefields	01 Permanent snow and icefields

Table 3.2 An extension of the Anderson *et al* (1972) level scheme to suit the New Zealand agricultural study based on Landsat data (Chapter 14).

Level I	Level II	Level III	Level IV	Level V
02 Agricultural land	01 Cropland and pasture	01 Cruciferae	01 Turnips	
			02 Kale	
		02 Cereals	01 Wheat	01 Spring wheat
				02 Autumn wheat
			02 Barley	01 Spring barley
			03 Oats	01 Oats for feed
				02 Oats for seed
		03 Legumes	01 Peas	01 Spring peas
				02 Autumn peas
		04 Pasture crop	01 Lucerne	01 Mown
				02 Standing
			02 White clover	01 Not flowering
				02 White heads
				03 Brown heads
			03 Rape seed	
		05 Pasture grazed	01 Heavily grazed	01 White clover/ryegrass
				02 Lucerne
			02 Lightly grazed	01 Clover/ryegrass
				02 Lucerne
02 Unproductive land		01 Bare ground		

input other than data from the sensors alone, e.g. topographic map/ recreational map material. Such a combination could improve the level of land use classification. They remark that advances in technology will also support further refinements in classification levels.

This two-level land cover classification scheme is presented in table 3.1. Level I is regarded as being directly recognisable from photographic data products prepared from the satellite imagery. Recognition is aided if these products are colour photographic images as opposed to black and white products. Level II would usually require either a more detailed examination of larger scale (1:100 000 as opposed to 1:1 000 000) imagery and/or the review of applicable subsidiary supportive information, e.g. topographic maps.

Class 02-01, the cropland and pasture subdivision of agricultural land (Anderson *et al* 1972) is the major Level II class investigated in the agricultural classification studies (Chapter 14). This broad Level II classification is not sufficiently detailed to support the classifications reported here. However, following the guidelines of Anderson *et al* (1972) the concept can be easily extended. Class 02-01 is thus divided into subclasses to Level V (table 3.2). These higher level groups were further subdivided and used in the agricultural study presented in Chapter 14. This was made possible by utilising imagery with greater spectral and spatial resolution, from an eleven-channel aircraft scanner, recorded at an opportune time in the cereal growing season—mid-summer.

3.3 References

Anderson J R, Hardy E E and Roach J T 1972 *A land-use classification scheme for use with remote sensor data* USGS Circular No 671 (Washington, DC: US Geological Survey)

<table>
<tr><td>

4

</td><td>

An Introduction to the Landsat Multispectral Scanner

</td></tr>
</table>

4.1 The Landsat Programme

The Earth Resources Technology Satellite (ERTS) Programme was established by the United States National Aeronautics and Space Administration (NASA). The ERTS spacecraft, later renamed Landsat, are unmanned satellites recording radiances of earth features.

Landsat was

'designated as a research and development tool to demonstrate that remote sensing from space is a feasible and practical approach to efficient management of the earth's resources' (NASA 1971).

The mission of Landsat is to provide a satellite for the repetitive acquisition of high resolution multispectral data of the earth's surface. These

'data products will be used by investigators for developing practical applications in the various earth resources disciplines including agriculture, forestry, geology, geography, hydrology and oceanography' (NASA 1971).

4.2 Observatory System and Payloads

The elements of the Observatory System are shown in figure 4.1. The payload on board comprised two remote sensor systems (MSS and RBV), Wide Band Videotape Recorders to record the data and the spaceborne portion of a Data Collection System (DCS). The data that form the basis for the analysis projects reported here were acquired by the MSS system.

4.2.1 Multispectral scanner (MSS)

This was the major sensor on board the Landsat satellites and is retained aboard the new series commencing with Landsat 4. The MSS is a line-scanning device which uses an oscillating mirror to continuously scan in a direction perpendicular to the spacecraft's velocity. The combination of the oscillating mirror and spacecraft progression for Landsat 1, 2, 3

15

scanned an area of 185 × 185 km² with an effective ground resolution of 56 m (cross track) by 79 m (along track) (USGS 1979). These resolution elements are situated on the easternmost edges of the Instantaneous Field of View (IFOV) accepted by the Landsat scanner's optical system—as indicated in figure 4.2.

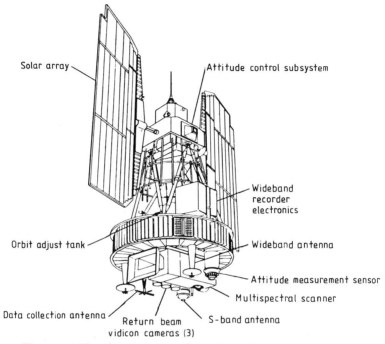

Figure 4.1 The observatory configuration of the Landsat satellite.

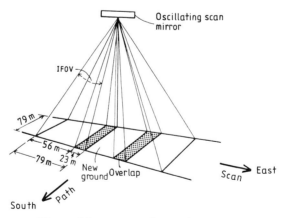

Figure 4.2 MSS ground scanning pattern.

The MSS system generally recorded in four different spectral bands as shown in table 4.1 (a provision did exist in some cases for a thermal channel but data were never readily available).

Table 4.1 MSS spectral bands.

Band	Spectral interval	Spectral range (μm)
Band 4	Green	0.5–0.6
Band 5	Red	0.6–0.7
Band 6	Infrared	0.7–0.8
Band 7	Infrared	0.8–1.1

For the data received on tape in New Zealand, each band of each image set usually consisted of 2340 scan lines (perpendicular to the orbital track) and 3264 picture elements (pixels) along each scan line. Each pixel had its radiance 6-bit digitised (i.e. $2^6 = 64$ possible levels) and these were decompressed to cover the actual range 0–127 for the radiance level for MSS 4, 5, 6 but were retained as linear for the range of MSS 7, 0–63 (USGS 1979).

4.3 Orbit and Coverage

The Landsat satellites revolve around the earth in a circular, sun-synchronous, near-polar orbit at an altitude of 880–930 km with a nominal 0930 local time crossing of the equator on the southbound recording pass. The satellite circles the earth every 103 minutes with 14 orbits per day. This provides an 18-day repeat coverage at the same local time, such that image centres are repetitive to within 37 km (USGS 1979).

4.4 Path for Acquiring Recorded Data

Prerecorded data were transmitted directly from the satellite observatory to a ground receiving station in the United States (either Gilmore Creek in Alaska, Goldstone, California, or Goddard Space Flight Center outside Washington, DC) once the spacecraft was within range (about 2400 km) of such a station.

The initial recordings of the data at the ground stations were then sent to Goddard Space Flight Center where the data were processed, with the housekeeping information, and stored on high density tapes (HDT). These latter tapes were then shipped to the Earth Resources Observation

Systems (EROS) Data Centre (EDC) in Sioux Falls, South Dakota. Here the Computer Compatible Tapes (CCT) are prepared from the HDT. These CCT are then forwarded to users. Such CCT can be treated as input data for analysis projects.

4.5 References

NASA 1971 *Landsat Data Users Handbook, Document No* 71SDS4249 (Goddard Space Flight Center, MD: NASA)
USGS 1979 *Landsat Data Users Handbook—Revised Edition* (Arlington, VA: US Geological Survey)

5 | An Introduction to an Aircraft Multiband Scanner System

5.1 The Aircraft Multiband Scanner (AMS)

Over the period 3 December 1980 to 11 January 1981, 39 project sites were overflown in New Zealand and eleven-channel digital imagery was obtained from an Aircraft Multiband Scanner (AMS) hired from Land Resources Management Ltd of California, USA. This was the basis for the New Zealand digital AMS database. Some of these projects were subsequently analysed within the framework of the IBM Joint Research Program Agreement using the ERMAN system in Sydney, Australia. As such, these projects were chosen to follow on from the Landsat project and thus experience with higher resolution (spectral and spatial) data was gained in preparation for the Landsat Thematic Mapper and SPOT systems.

Here we will be introducing the AMS concept. Exact details of any aircraft scanner system will vary between units but the basic concepts will be common.

The AMS was used to acquire the data reported in Chapter 15. It is a modified version of a Bendix scanner (Bendix Aerospace Systems Division 1974, Land Resources Management 1980).

5.2 The AMS Instrument

5.2.1 Spectral bands

The AMS is an eleven-channel digital multispectral scanner operated from an aircraft platform at typical altitudes from 400 to 4000 m. The eleven channels of this AMS spanned the visible and near infrared wavelengths and one channel operated in the thermal infrared (table 5.1). The aircraft scanner covered a wider range of wavelengths, with narrower channels, than did the Landsat scanner (figure 5.1).

Table 5.1 Channel allocations for the AMS cited here

	Channel (band)	Wavelength (μm)	
		Centre	Bandwidth
Visible	1	0.410	0.060
	2	0.465	0.050
	3	0.515	0.050
	4	0.560	0.040
	5	0.600	0.040
	6	0.640	0.040
	7	0.680	0.040
	8	0.720	0.040
Near infrared	9	0.815	0.090
	10	1.015	0.090
Thermal	11	11.0	2.5

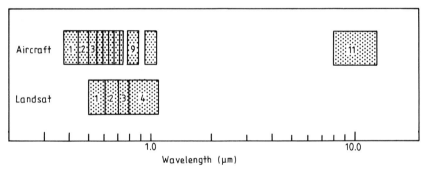

Figure 5.1 Comparison of the aircraft scanner reported in this chapter and Landsat MSS wavelength bands.

5.2.2 AMS *scan geometry*

The size of the Ground Resolution Element (GRE) immediately below the aircraft (at nadir) is directly related to the Instantaneous Field Of View (IFOV) and the flying height (h), as follows:

$$(\text{GRE})_N = h(\text{IFOV}) \tag{5.1}$$

where $(\text{GRE})_N$ indicates the nadir GRE is being considered.

For the AMS used as an illustration here, the IFOV was 2.5 mrad. The nadir pixel sizes for various flying heights are presented in table 5.2.

This GRE enlarges as the scan angle θ increases away from the nadir. To a first approximation, as the IFOV is much less then θ, the situation may be portrayed as in figure 5.2.

Table 5.2 The relationship between ground resolution and aircraft height, and the selected scan rates appropriate for varying ground speeds.

Height (m)	Nadir pixel size (m)	Ground speed (knots)								
		100	110	120	130	140	150	160	170	180
400	1	52	57	62	67	72	77	82	88	93
600	1.5	34	38	41	45	48	52	55	58	62
800	2	26	28	31	33	36	39	41	44	46
1000	2.5	21	23	25	27	29	31	33	35	37
1200	3	17	19	21	22	24	26	27	29	31
1400	3.5	15	16	18	19	21	22	24	25	26
1600	4	13	14	15	17	18	19	21	22	23
1800	4.5	11	13	14	15	16	17	18	19	21
2000	5	10	11	12	13	14	15	16	18	19
2200	5.5	9	10	11	12	13	14	15	16	17
2400	6	9	9	10	11	12	13	14	15	15
2600	6.5	8	9	10	10	11	12	13	13	14
2800	7	7	8	9	10	10	11	12	13	13
3000	7.5	7	8	8	9	10	10	11	12	12
3200	8	6	7	8	8	9	10	10	11	12
3400	8.5	6	7	7	8	8	9	10	10	11
3600	9	6	6	7	7	8	9	9	10	10
3800	9.5	5	6	7	7	8	8	9	9	10
4000	10	5	6	6	7	7	8	8	9	9

Scan rates (scans per second)

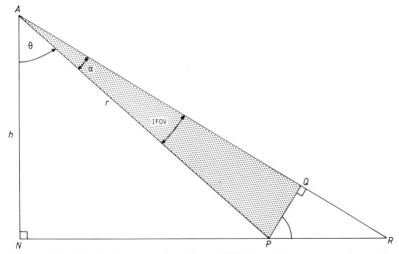

Figure 5.2 Relation between the Aircraft Multiband Scanner A, the nadir position N, and the Ground Resolution Element (GRE) at scan angle θ for a set Instantaneous Field of View (IFOV) and flying height h.

If h is the flying height then r is given by

$$r = h/\cos \theta \qquad (5.2)$$

In figure 5.2 we are looking along the flight direction with the scan taking place out to the side (towards P from N). The GRE dimension along the flight direction $(\text{GRE})_F$ is thus given by equation (5.3)—a modification of (5.1) by (5.2):

$$(\text{GRE})_F = h(\text{IFOV})/\cos \theta. \qquad (5.3)$$

A different situation applies in the scan direction. Here the assumed square pixel dimension, which is based upon a square entrance aperture series of field steps, must be mapped from PQ to PR. This pixel dimension along the scan direction is designated by $(\text{GRE})_S$ and is derived as follows:

$$(\text{GRE})_S = \text{PR} = \text{PQ}/\cos \theta$$

and

$$\text{PQ} = (\text{GRE})_F = h(\text{IFOV})/\cos \theta.$$

Hence

$$(\text{GRE})_S = h(\text{IFOV})/\cos^2\theta. \qquad (5.4)$$

Thus for the agricultural project studied using this AMS, and discussed in Chapter 15, in order to have a selected nadir pixel size of 10 m, the GRE dimensions at a scan angle of 45° will be, for a flying height of 4000 m,

$$(\text{GRE})_F = 14.14 \text{ m} \qquad (\text{GRE})_S = 20.00 \text{ m.}$$

5.2.3 AMS scan width, length

Like the GRE the scan width is also a function of flying height. For a scan of ±57.5° a ground swath of 8.03 km could be imaged from 4000 m. However, from practical considerations, if we limit the usable data range to scan angles between −45° and +45°, as the $(\text{GRE})_S$ has doubled at that point over the requested nadir GRE, the usable imaged swath is then 6.28 km.

The scan length is purely a function of the available storage medium, flying conditions and project parameters.

5.2.4 AMS scan rate

The AMS instrument discussed here is capable of operating at between 10 and 100 scans per second. The desired scan rate is determined by the required nadir pixel size, $(\text{GRE})_N$, and the aircraft speed.

For light aircraft, such as used for the project reported here, a useful speed for stability is 140 knots. At this speed the aircraft advances along the nadir track at 72.02 ms^{-1} (based on 1 knot = 0.514 44 ms^{-1}). If the desired nadir pixel size is 10 m then a scan rate of 7.20 scans per second is theoretically required for contiguous nadir pixels. In practice some along-track overlap is desirable to account for slight pitch irregularities in the aircraft motion, non-uniform response over the IFOV, etc. Using this approach the scan rates for a variety of altitudes and aircraft speeds are presented in table 5.2—the above example pertained to the agricultural case discussed in Chapter 15.

With the lowest available scan rate being 10 scans/second the approach for the higher altitude runs is/was to use double scanning. Thus for the above example the scan rate would have been chosen as 14 scans/second. This redundancy in the data can safeguard against flight (pitch, roll, yaw) irregularities.

5.2.5 Optics and detectors

The optics of the AMS instrument discussed here include reflective scanning and focusing optics, a ten-channel spectrometer and a thermal infrared detector. A single-sided 45° mirror rotates to provide lateral scanning of the earth's surface through an elongated window in the underside of the aircraft.

The scan window was 120° but only 57.5° either side of the nadir generated reliable data. The additional 5° permitted compensation for aircraft roll. The scanner optics provided an IFOV of 2.5 mrad.

5.2.6 Data recording

The analogue video signals from eleven channels were 8-bit (0–255 range) digitised and recorded aboard the aircraft on a High Density Digital Taperecorder. For the AMS used here, 803 samples were taken along each scan line, centred on the nadir track over all eleven channels. Upon return to ground these data tapes were transcribed to Computer Compatible Tapes (CCT) for subsequent processing on usual computing systems.

5.3 Correction of the AMS Imagery

The mode of operation of the airborne multispectral scanner causes inherent geometric and radiometric errors in the imagery. During the processing of the CCT which were the results of the 1980/81 summer scanner survey, it became apparent that the usefulness of the imagery could be improved by removing some of these distortions in the raw data. Two corrections are considered to be of particular importance: the removal of 'S-bend' distortion, and the removal of radiance level shifts caused by imperfect scanner calibration control. This section describes these two corrections and also indicates other degradations which should be borne in mind when planning an aircraft scanner programe.

5.3.1 S-bend correction

The compression of the aircraft scanner data at either end of the scan line due to increasing pixel size combined with the along-track movement of the aircraft during each scan results in an 'S-bend' distortion, when reproduced as an image. The distortion is exaggerated by scanning over a relatively wide scan angle.

To eliminate this distortion the following technique was adopted:

(i)   the scan line image on the ground was divided up into a fixed number of pixels of equal length in the direction of the scan;

(ii)   the new pixel size was made equal to the nadir pixel size;

(iii)   each scanner pixel was located at its true position in this new grid on the basis of its scan angle;

(iv)   gaps in the grid were filled by duplicating the nearest pixel. (Similarly an interpolative procedure could have been employed but the faster nearest neighbour approach was chosen in this instance.)

The original data from the scanner contained 803 pixels corresponding to a scan angle of $\pm 57.5°$ from nadir (the nadir pixel being pixel number 402). In order to avoid the low resolution data at the edges of the scan, planning for all surveys was done on the basis of $\pm 45°$ of usable data. To retain the greatest compatibility with the original data a standard corrected grid of 803 pixels was chosen (± 401 pixels on either side of nadir). The scanner viewing geometry is as shown in figure 5.2.

The position of a pixel in the corrected grid is calculated as follows. If θ is the scan angle from nadir to the pixel of interest, h is the aircraft flying height above ground and d is the dimension of the nadir pixel, then from figure 5.2

$$NP = h \tan \theta$$

and hence

$$P = h \tan \theta / d$$

where P is the new pixel number (from nadir) in the corrected grid.

Also, at nadir

$$h = d/(\text{IFOV}).$$

Therefore

$$P = \tan \theta /(\text{IFOV}).$$

Now accepting that θ is made up from contiguous IFOV in the original data we may write

$$\theta = n(\text{IFOV})$$

where n ranges on either side of the nadir from 1 to 401. n is the number of the pixel in the scan line displaced in either direction from the nadir pixel.

Consequently the relation between the number of the pixel in the recorded scan line dataset and in the S-bend-distortion-corrected scan line dataset is given by

$$P = \tan(n\alpha)/\alpha \qquad \text{for } n = 1, \ldots, 401 \qquad (5.5)$$

where α is the IFOV angle and n is the number of the pixel displaced from the nadir pixel.

It should be noted that within the scan angle of interest ($\pm 45°$), no pixel in the corrected grid is displaced by more than one pixel from its original neighbours (i.e. gaps in the corrected grid that require to be filled by pixel duplication are no more than one pixel wide).

5.3.2 Calibration correction

During normal operation of such a scanner, a predetermined range of input signal level is selected to correspond with the dynamic range of the analogue to digital converter. This is accomplished by examining the analogue signal while the scanner is looking at high and low level calibration sources. The low level calibration signal is adjusted by means of an offset to equal the preselected low level reference, and the gain is adjusted to bring the high level calibration signal down to the preselected high level reference.

The data from the calibration sources are usually recorded on the CCT as part of a header data block associated with each scan line. Any variations in the high and low level calibration sources can be used to correct the radiance data for the scan line pixels. To apply such a correction, reference calibration levels need to be determined at a point in the survey where the scanner was known to be operating correctly. The image can then be processed by introducing new offset and gain corrections such that the calibration levels for each line are restored to the reference levels.

Because the signal level on these AMS data was at times significantly greater than the high calibration level, there was a tendency for small fluctuations in calibration level to be exaggerated by the scaling factor during gain correction. It was assumed that small variations on a line by line basis were probably caused by system noise, rather than real changes in gain. To remove this effect, the calibration levels were recursively filtered prior to the corrections being applied. It was found that if filtering was not applied the calibrated image was significantly noisier than the original. This noise level is subjectively more apparent than the overall calibration errors.

The offset used was calculated from the low calibration level as follows: for each new scanner line, the offset was set equal to the average of the previous offset and the low calibration level. The required gain was taken to be the ratio of the new to the original calibration level range. The new calibration level range was the average of the previous calibration level range and the difference between the high and low calibration levels for the new line. This degree of recursive filtering seemed most satisfactory.

5.3.3 Other image degradations

The following additional spatial and spectral degradations can also be present in AMS data. They are outlined here solely for user awareness since they should be borne in mind when planning an AMS survey.

5.3.3.1 Spatial distortions from ground height variations

A change in pixel size occurs in areas of undulating terrain. The nadir pixel size is a function of the height of the aircraft only, as the instantaneous field of view of the scanner is fixed. Thus, if there is a 1% change in height of the aircraft above the ground, there will be a corresponding 1% change in the pixel size.

A more important effect of ground height variations, even in flat areas, is pixel displacement due to parallax. A 30 m change in altitude of the land at a 45° scan angle will result in a three-pixel displacement in the data.

The implication of this ground height variation is that a mean ground level must be chosen as part of the flight planning procedure. In general, it is not desirable to vary aircraft altitude during a given project to accommodate ground height variation.

5.3.3.2 Spatial distortions from aircraft instability

This is a potentially serious problem since large amounts of data can be lost because of relatively small aircraft movements. A change in aircraft altitude changes pixel size in the same manner as variation in ground height (§ 5.3.3.1). Changes in pitch, roll and yaw angles also produce spatial distortions in the data.

A change in pitch angle from the nominal will result in along-track distortion. The extent of this will depend upon the rate of the pitch and the AMS scan rate. Yaw angle variation results in a rotation of the scan line direction about the nadir point and is greatest at the end of scan lines, and least at the centre. Again the extent of this variation will depend upon its rate with respect to the scan rate. The scan mirror is usually roll-stabilised in AMS units so that roll distortions should not occur in the data.

While pitch and yaw changes can produce significant loss of data, this loss is not immediately obvious in the photographic product. The wisest choice is to fly with experienced AMS operator/pilot teams under stable air conditions.

5.3.3.3 Radiometric degradations

A 'radiance fall-off' occurs along the scan line from the nadir out to the edge of the scan due to the increase in ground to scanner pathway. The longer the pathway, the more radiation attenuation and scattering can occur because of the atmosphere.

A potentially more serious problem is variation in brightness, which occurs across the scene if the flight line is not directly towards, or away from, the sun. It is not always possible to fly a project under optimal sun angle conditions. This problem is particularly significant in study areas which include bodies of water. In this situation specular reflections may overload the scanner.

5.4 Project Planning

From the foregoing discussions, a number of factors have emerged that must be considered when planning an AMS survey. Some of these affect the desired resolution and others need to be guarded against to preserve data fidelity. Some of the major flight objectives when planning a survey are now outlined.

5.4.1 Resolution

Resolution is one of the most important factors in planning a project. Once this has been chosen, aircraft altitude and scan width are fixed as the aircraft is flown at a height appropriate to the desired data resolution. The resolution varies from the nominal, directly beneath the aircraft, to half the nominal at a 45° scan angle. This affects the choice of resolution for a project.

Further, the spectral contrast between the target and its background must be considered in determining the required project resolution. At maximum scan angle, the ground resolution element should be no greater than half the target size. For targets of low spectral contrast, the ground resolution should be selected to be one third to one fifth of the target size.

5.4.2 Altitude

If the flying height is not determined by the required resolution, it should be selected to match the width of the project area, remembering that distortion increases and resolution decreases towards the edges of the scan. If significant changes in ground height occur, a mean ground level must be chosen for the project.

5.4.3 Flight direction

If the choice is available, the flight line should be directed towards or away from the sun to avoid variation in brightness across the scan line (see § 5.3.3.3).

5.4.4 Time of flight

Projects should usually be flown at high sun angle (i.e. a couple of hours either side of midday) to minimise sun angle effects. Where possible, NE/SW flight lines should be flown in the morning and NW/SE lines in the afternoon in the Southern Hemisphere and the reverse for the Northern Hemisphere. However, in some cases the time of flight is determined by the nature of the project. For instance, frost studies and geothermal studies which utilise the thermal band information must be flown before sunrise.

5.4.5 Other considerations

Clear, calm flying conditions are needed for good scanner imagery. Minor turbulence can cause severe distortions in the imagery and clouds can either obscure the target area or cast shadows within it. Only the thermal channel (band 11) may be unaffected by cloud shadows.

5.5 References

Bendix Aerospace Systems Division 1974 *Modular Multispectral Scanner (M²S)* **BSD** 8039 (Ann Arbor, MI: Bendix Aerospace)
Land Resources Management Inc. 1980 *M²S Modular Multiband Scanner, HDDT Interface* (California: Land Resources Management)

6 Introduction to Image Enhancement and Analysis

6.1 Outline of Image Enhancement and Analysis

Computerised image processing uses the digital data and falls into two main streams: image enhancement and image analysis. The former covers: destriping, haze correction, linear and non-linear grey scale or colour (hue) enhancement and textural enhancement. The result is usually a photographic product in which the features of interest can be more clearly distinguished by a human interpreter. The latter, image analysis, leads to computerised classification of like ground cover classes and their display on line printer and photographic products. This chapter provides an introduction to the major enhancement and analysis procedures that are currently available for users to interact with their digital image data.

The Landsat data discussed here, as example data for these processes, were taken from the Computer Compatible Tape (CCT) product provided by the Earth Resources Observation Systems Data Center in Sioux Falls, South Dakota, USA.

6.2 Major Image Enhancement Processes

6.2.1 Destriping

Each Landsat multispectral band is imaged onto a bank of six detectors (USGS 1979). This gives a total of 24 detectors covering the four Landsat bands. As each detector has different characteristics the incoming radiance can be scaled to different digital levels between the detectors. These characteristics may be a combination of zero level offset, gain, linearity of the gain over the dynamic range and the dynamic range itself. As a result, the contiguous scan lines recorded over the same ground target by adjacent detectors can differ slightly. This can be up to ± 5 CCT levels from the mean. This difference in the recorded radiances can lead to obvious scan line striping of the image. The effect is exaggerated by textural enhancement, and also by colour enhancement if the enhancement is set up for a narrow radiance range. The striping can also lead to the corruption of classification results if the classes are tightly defined and tend towards overlap in spectral space.

Horn and Woodham (1979) outline one of the most effective current techniques for overcoming the above problems. It compensates for any non-linearities in the gain over the dynamic range of each detector. Small variations in each detector's characteristics can also be adjusted for, provided that the differences in the dynamic range are not too extreme.

The technique relies upon the assumption that the same ground cover class distribution will be sensed by each detector with an approximately equal probablility provided that the sampling area is large enough. Once the six radiance (CCT level) occurrence distributions are compiled, one for each of the six detectors, a total occurrence distribution is prepared. The 'peaks' in the *individual* detector's distributions are compared with the 'peaks' in the *total* distribution. Then a modified look-up table is prepared to translate the data 'peak' in the individual detector's distribution in order to match the 'peak' in the total distribution. This technique is highly responsive to any system non-linearities and is usually felt to be the most applicable.

6.2.2 Haze correction

The Landsat spacecraft uses passive sensors to record data on ground cover classes. This means that the sun is the source of irradiating energy, and the reflected energy is received, filtered and used to produce a digital signal at the spacecraft. As the solar irradiance passes into and through the atmosphere a certain amount will be reflected and re-radiated back in the direction of the satellite. Similarly some of the radiant energy which reflects from the ground is absorbed during the passage through the atmosphere to the sensors in the satellite. A simple correction cannot be applied for the absorption but a correction can be included for the atmospheric reflection/scattering/radiation component of the signal reflected from the ground. As the atmospheric scattering will constitute a base level to all radiances received from valid ground cover classes a subtraction of this 'zero offset' can be applied to all recorded radiances, as a first-order correction. That is, if the first CCT level that has a non-zero occurrence in the occurrence level histogram is a, then all radiances may be shifted downwards by a to compensate, to a first approximation, for this atmospheric scattering. This is known as 'haze correction' (USGS 1979).

6.2.3 Linear stretching

Original Landsat data on a CCT are usually concentrated within a limited subset of the available range of levels (0–127 for MSS channels 4, 5, 6, and 0–63 for MSS 7—see Thomas (1973) for a background discussion).

If an occurrence histogram were prepared from the original CCT data, as a function of the CCT levels, an abundance of major and minor peaks would usually be seen towards the lower end of the overall CCT level range in each channel. Without enhancement these distributions generally produce dark and low contrast images.

A simple form of grey scale enhancement, the linear stretch (USGS 1979), can be quickly applied to the original data and often produces more than satisfactory results. To implement the linear stretch, the radiance range that describes either the whole image or a particular area of ground cover is first found from the occurrence distribution limits. Having established the set of upper and lower limits the original data are stretched between these limits to fill the radiance range of the output device.

Linear stretching has the advantage over some non-linear stretching techniques in that the minor occurrence peaks are preserved rather than being absorbed into the major peaks. Consequently it may be possible to better differentiate those species that occur infrequently on the basis of colour difference using this technique, rather than by using some non-linear stretching processes.

Linear stretching may also be applied on a segment-by-segment basis throughout an image. This permits more appropriate contrast enhancement for, say, a block of forest alongside a lake which may then be adjacent to farmland. The image would be broken into the major segments and individual linear stretches determined. These would then be applied, via access to different look-up tables, as the appropriate enhancement treatment is requested for each pixel in the image. This technique is outlined by Fahnestock and Schowengerdt (1983).

A development from this regional technique could lead to an even more 'purpose built' enhanced product for the user. The user may be particularly interested in displaying the maximum details in the data for a limited range of broad classes: e.g. forests, pastoral land, water; and be content with a more general enhancement for the remainder of the image. To achieve this 'class-specific' enhancement, a classification of the image into these broad classes could be run first. This would produce both a single-channel classification mask (i.e. a pixel would be labelled as being one of the forestry, pastoral, water, 'unclassified' categories) and a set of statistics for each class. The appropriate look-up tables for enhancing the various classes (here forestry, pastoral, water) would now be prepared, one for each class for each of the desired enhancement output channels, from the class and channel statistical data. A general image enhancement look-up table would also be prepared. The complete image would now be enhanced for output as follows: as each pixel is entered into the enhancement phase, its classification status (e.g. forestry, pastoral, water, 'unclassified') would be evaluated. If it were one of the sought classes (here forestry, pastoral, water) the look-up table appropriate to that class and that channel would be entered. This would yield the more 'class-specific' enhanced value for the pixel being enhanced. If the pixel fell into the 'unclassified' class then the general enhancement look-up table would be applied. The alternative, and usual, approach is to take the statistics for the complete image and apply the resultant less effective, in this example, treatment to the whole image.

6.2.4 Non-linear stretching

Non-linear stretching is confined here to the histogram equalisation process that is available, usually routinely, in image processing software suites. There are other forms of non-linear stretching based on the occurrence distributions but all essentially follow the same broad derivation logic.

Each CCT level usually has a different number of occurrences associated with it. If the 'area' of each occurrence block (abscissa × ordinate) is scaled such that the available range 0–255, say, is divided proportionally on the basis of those areas, we have the process known as histogram equalisation. Obviously the major peaks will occupy a greater number of radiance levels than they did in the original data distribution. Similarly the minor peaks could occupy less, or be included in the range of levels allocated to the species that typify the major peaks. Some separability of less frequently occurring classes can be lost in this process. However, the process does enable a visual interpreter to most easily differentiate the major species on the basis of colour, or grey scale, separability. The user must weigh the advantages and disadvantages of either a linear or a non-linear stretching approach. A careful choice of class training fields can offset the disadvantages of either technique.

Any form of colour enhancement (linear/non-linear stretching) applied to striped data can exaggerate the striping if the target region is small and the overall number of ground data radiance levels is low. The striping will then produce small CCT level departures from the ideal and, whilst small in CCT level terms, this could be significant in occurrence level terms. As a result the striping departures can become pseudo-data classes. This is particularly important if the colour enhancement modules are applied to texturally enhanced datasets that have both *not* been destriped and contain few true ground radiance levels. Any striping departures can be unwarrantably emphasised by this process.

6.2.5 Textural enhancement

The colour enhancements described above are appropriate to maximising the *colour* separation between homogeneous regions. Textural enhancement emphasises the boundaries between such regions.

A variety of textural enhancement algorithms are in use. All rely essentially on referring the radiance level for a 'central' pixel in an $n \times m$ matrix to the 'average' level of those surrounding it. A replacement level is then determined that exaggerates the departure from the average. This revised level is then stored in the output dataset and the process is repeated using the input data from the next pixel, and so on. Two common approaches are subtractive box filtering and Laplacian textural enhancement.

Subtractive box filtering uses the following relationship:

$$R^1 = R - (F \times \bar{R}) + C$$

where R is the radiance in the selected MSS band, from the pixel being processed; $\bar{R}$ is the average radiance from the $n \times m$ nearest neighbour matrix surrounding, but excluding, this pixel; R^1 is the synthesised radiance which replaces R in the output dataset; F is the fraction of the average surrounding radiance that was taken to be the base level; and C is an additive constant.

As a starting point the values of $F = 0.8$ and $C = 20.0$ can be used and have usually proved effective for 3×3 matrices (Thomas *et al* 1979).

Laplacian textural enhancement involves the convolution of a filter matrix with the pixel radiance matrix. It produces a revised output radiance, stored in the output dataset, of R^1.

If the 3×3 pixel radiance matrix is taken to be

$$\begin{bmatrix} R11 & R12 & R13 \\ R21 & R22 & R23 \\ R31 & R32 & R33 \end{bmatrix}$$

and the filter is

$$\begin{bmatrix} 0 & -1 & 0 \\ -1 & +5 & -1 \\ 0 & -1 & 0 \end{bmatrix}$$

then $R^1 = -R12 - R21 + 5 \times R22 - R23 - R32$.

The above treatments are non-directional and give good overall textural enhancements. Other filter matrices can be easily fabricated, both in the number of dimensions and in the values, to give different levels of general edge enhancements or directional enhancements (see for example Bernstein 1978, Hall and Awtrey 1980).

Enhancement for edges alone may be easily effected using the above Laplacian filter. If the $+5$ is replaced by $+4$ as the central element in the filter only the edges will be seen. The addition of the original single-channel image data is accomplished using the $+5$ as the central element.

Convolving the filter and data matrices exaggerates any changes of radiance that occur, and leaves untouched those areas in which the radiance levels are all the same. The overall effect is to 'sharpen' all boundaries, and to increase the apparent resolution.

A problem can arise if a previously texturally enhanced dataset is then histogram equalised for colour enhancement. The textural enhancement will usually introduce at least two further peaks into the occurrence histogram. These peaks are produced by the 'edge effect' pixels, which are raised or lowered with respect to the 'homogeneous' core region. Histogram equalisation is usually set up to enhance a limited range of ground cover classes. Consequently the portion of the histogram for each ground cover class will have a central data peak flanked by the 'edge' peaks. This produces an occurrence histogram for the enhancement region tending to less distinct data peaks than occurred in the raw data—the 'valleys' are being filled. As a result, a more continuous monotonically increasing look-up table will be produced for the texturally enhanced data than for the 'original' data. This trend towards a monotonic increase, over the three channels, will lead to more equivalent light levels being passed to the film through the blue, green and red filters for a colour product than would have occurred if the look-up table had been set up from the 'original' data. More unsaturated colours will thus be produced from the texturally enhanced data than from an enhancement of the 'original' data. Depending upon the magnitude of the unsaturation, a trend towards a black to white grey scale will be observed in the resultant colours in the enhanced product. The combination of these influences can lead to the histogram equalisation process producing an overall compression of the true data peaks in colour space and a tendency towards less saturated colours. That is, the resultant image can look 'flat' and tend towards white. A solution to this problem is to prepare the histogram equalisation look-up file from the basic data and run the texturally enhanced dataset through that look-up file. (No additive or multiplicative constants should have been used during the textural enhancement process.)

Obviously textural enhancement should only be applied to datasets that either have minimal striping or have been passed through the destriping process.

6.3 Major Image Analysis Processes

Image analysis is here taken as that series of processes that reduces the raw digital data to a 'map' display of ground cover classes, together with a summary of the occurrence statistics for each class over the test area being considered.

Fundamental to such a classification process is the derivation of a spectral signature for each ground cover class. For a supervised classification this spectral signature is extracted from the digital data by the analyst and entered into the computer classification module. (In this way it is a 'supervised' process. That is, the analyst trains the classification process to those characteristics he requires to be considered. 'Clustering' permits the computer to aggregate classes into the most frequently occurring groups. Clustering is unsupervised and less able to differentiate classes that are similar than the supervised classification process. We shall here confine ourselves to the more user-interactive classification techniques.)

The spectral signature for a ground cover class is the 'quantitative measurement of the properties of a (class) at one or several wavelength intervals' (Reeves 1975). Using Landsat we have, usually, access to the basic four data bands (MSS 4, 5, 6, 7) and sometimes access to other synthetic channels generated from these bands (see § 6.5).

6.3.1 A basic classifier

The simplest form of classifier is the supervised histogram parallelepiped classification. This is henceforth abbreviated to parallelepiped classification. It relies on simplified spectral signatures being inserted into the software for each data band. These simplified spectral signatures are just lower and upper limits to the occurrence histogram for each target class in each data band. The exact setting of the limits along the occurrence histogram abscissa (CCT level axis) is done by the user—a human (supervised) decision. Once the lower and upper limits have been defined for one channel they form a dimension for one side of a parallelepiped. When the second set of limits, for the second channel, is defined, so too is a two-dimensional parallelepiped. Similarly, extending the discussion to four channels gives a four-dimensional parallelepiped.

In operation, a pixel possesses four CCT levels that describe its position in four-channel, or four-dimensional, feature space. If this position is within the four-dimensional parallelepiped, for a particular class, then the pixel is regarded as reflecting the characteristics of that class—within the constraints of the spectral and spatial sampling of Landsat. This means that for MSS 4, the pixel's MSS 4 radiance value must be between the lower and upper limits selected for that class, or be equal to them. The same constraint must also be true for the other three channels.

6.3.2 Production of a thematic map

To produce a thematic map a number of steps must be undertaken:
    (i)   spectral signatures must be determined;
    (ii)   a classifier must be run;
    (iii)   statistics must be obtained;
    (iv)   thematic maps must be produced on the lineprinter as a character-coded product or on magnetic tape as numbers signifying colour codes;
    (v)   the colour-coded thematic map has to be outputted to colour transparency film;
    (vi)   a form of spatial postprocessing may be advantageous;

(vii) a method of relating the colour-coded themes to their physical location on the ground must be provided.

Let us look at these steps in order.

6.3.2.1 Determination of spectral signatures

The analyst must first define a homogeneous training field for the class under consideration. On a microscopic scale, the field may consist of a mixture of targets which, on the macroscopic scale, are defined as a single class. The training field must be a representative of that mixture class, and yet be sufficiently homogeneous to enable the class to be adequately identified.

Figure 6.1 (original source unknown) provides an illustration of this process. In the top left corner is portrayed an image of a face with fine

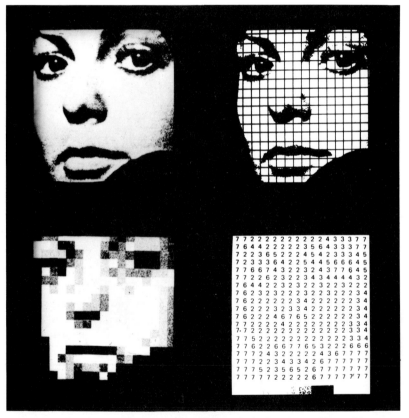

Figure 6.1 This image illustrates the effects of placing a sampling grid over a microscopically heterogeneous image and turning it into a macroscopically homogeneous digital data set—as done by Landsat. A loss of resolution is apparent and must be considered when setting up class training fields.

detail being evident. Upon the imposition of a sampling grid (top right image) a restriction on the level of detail that can be resolved within each sampling pixel becomes apparent. Once a representative grey level is assessed, homogeneously for each pixel, it is digitised. The results from this process are given in the bottom right image. If then the image is reconstituted from the sampled digital data the image in the bottom left is produced. The compromise is then obvious, on comparing the upper and lower left images, between detail that is 'microscopically' discernible by a higher resolution sampling system—the human eye—and that which is regarded as being 'macroscopically' homogeneous by the sampling system. From gauging the level of this compromise the relevant level of class discrimination may be assessed. This important concept must be understood by each user analyst before attempting to separate ground cover classes via any form of remotely sensed imagery.

There are several ways of extracting spectral signatures from Landsat data. The easiest is that of using an interactive graphics terminal, linked to a host computer, to locate suitable training fields. The boundaries to these training fields are then selected via the cursor, driven by the user over a grey (or colour) image of the data, and then the statistics for that training field (or fields) are computed and returned to the analyst for acceptance and use, or modification.

Another method is that of manually interpreting lineprinter products. This requires less sophisticated hardware (and software) and is more widely available. However, it is laborious. A hard copy, though, is available for use in the field. On the lineprinter a set of characters can be used to represent a range of CCT levels, with each character referring to a predetermined increment. The numerical level may then be related to the topographic features. An outline of the major ground features may be discerned on each band and hence the location of the training fields deduced.

Once the training field has been found on such character maps—one for each data channel—an occurrence distribution, in CCT level terms, can be deduced for each training field from the character maps. The occurrence statistics for each channel then lead to a spectral signature. Such a signature relies on a human decision of appropriate limits to set for the parallelepiped classifier (e.g. upper and lower distribution extremities; the intercepts, at 5% or 10%, of total occurrence levels; the mean $\pm$ 1, 2 or 3 standard deviations (assuming normal statistics) etc). The decision should be justifiable, consistent and repeatable.

6.3.2.2 *Classification and statistics compilation for target classes*
The simplest form of classification output is a character-coded thematic map produced on the lineprinter. The overall occurrence statistics can also

be easily prepared and outputted for each class. A revised spectral signature may also be suggested based on the overall class occurrence statistics.

By repeating this classification process, each time refining the spectral signatures, a representative set of statistics may be derived. So far the technique is applicable to all supervised classifiers (see Chapter 7 also) and leads to a stricter more homogeneous definition of the training data. The analyst must ensure that the training data are never refined to such an extent that they lose applicability to the actual amalgam of 'microscopic' classes that are desired to be represented as a homogeneous 'macroscopic' class.

For a parallelepiped classifier it is also useful to consider exploiting the order in which target species are classified. One class may interact with another class—as shown by the overlapping of the occurrence distributions in one or more data channels. If this overlap occurs in all data channels a more constrained approach to the spectral signature determination may be warranted. Alternatively, the technique of classifying one target class before another in a sequential process may be used. This facility is easily supported by the parallelepiped classifier, and is akin to an *a priori* weighting of classes (see Swain and Davis 1978), but should be used with care.

6.3.2.3 *Thematic map output and output to photographic film*

The lineprinter product suffers from differential X, Y distortion due to inter-line and character spacings on the printer. It also permits only a small area to be easily viewed. However, the scale of this product is ideal for field checking.

To support the classification of a wider area, a photographic representation can often be usefully employed. The classifier can often direct a classified dataset to magnetic tape for subsequent output as coloured themes on a photographic transparency. A class number to colour number modification module is used to translate the number associated with each class into a number that, when passed through a special look-up table, produces the pre-selected colour block for each class.

A problem with any such colour representation of classes arises when some isolated pixels of one colour are viewed against a block of one colour in one part of an image and against a block of another colour in another part of the image. The perceived colour of the isolated pixels can tend to change. With this change in perception the isolated pixels may be allocated to the incorrect class. For this reason only those colours that can be clearly and unequivocally discerned alone and amongst a crowd should be used. This usually restricts the choice to some nine colours plus black and white.

6.3.2.4 Inclusion of spatial postprocessing for datasets that have been classified on an individual pixel basis

Most computer classification techniques (e.g. histogram parallelepiped, minimum Euclidean distance, maximum likelihood—see Chapter 7) usually operate in spectral, rather than image, space. Using these techniques, each picture element (pixel) is classified into a target category without reference to its spatial neighbours.

Variability in the spectral distribution for the same 'training' target over an extended geographical area can lead to classification 'noise' being introduced into a final thematic map. This variability can be produced by the influences of differing soil type, soil moisture regime, wind pressure on vegetation with the resultant change in radiance, individual farming practices, etc.

Consequently, some form of spatial postprocessing is often desired to aggregate like-classified pixels together by 'filling in' the unclassified gaps between lesser aggregates and by rejecting or changing the ascribed class for pixels that have possibly been misclassified due to spectral noise. One technique is described further by Thomas (1980).

Such spatial postprocessing of spectrally classified data is based on the evaluation of a 'proximity function' for each pixel for each target class, in the initially classified dataset. The possibly revised classification status for each pixel is then passed, after this spatial postprocessing, to an output dataset. This technique has been used by the New Zealand group since 1977 to good effect in land cover, forestry and bathymetric classification projects (Ellis *et al* 1978).

Timmins (1981) concludes that its use, at times, may remove validly classified picture elements when dissected terrain is classified and also in regions where classes have differing heterogeneity. Its use should thus be weighed against the homogeneity of the class training fields and the size of the sampling pixel compared with the size of the regions that the user wishes to classify as homogeneous class blocks.

As an example of spatial postprocessing let us apply it to the Darfield agricultural region. Figure 6.2 is a lineprinter thematic map prepared using a parallelepiped classifier on four classes. These same four classes, with the same spectral signatures from the same dataset, were then subjected to spatial postprocessing. Figure 6.3 is the resultant thematic map. As reported by Thomas (1980), in this instance, the inclusion of spatial postprocessing reduced the area of 'noisy' targets by 33.6% for bare ground and 53.6% for Kopara wheat. ('Noisy' targets were taken to be those that had a broad occurrence distribution, in CCT level terms, signifying that the class consists of purer sub-classes, e.g. Kopara wheat at different heights, spacings, stress, development, etc. 'Quiet' targets were taken as those that had essentially a homogeneous spectral characteristic.)

A 'medium' noise target (lucerne or alfalfa) was found to have an area decrease of 24.1% whilst the 'quiet' target (exotic forest) experienced a 0.0% area decrease (as could be expected). These results tend to confirm the inferences that were drawn from Timmins (1981).

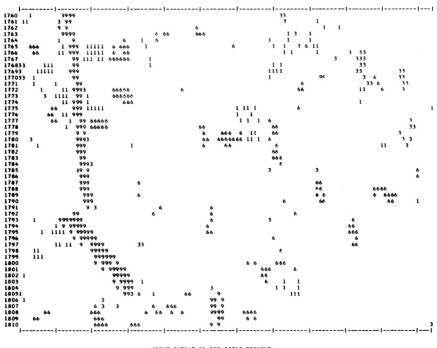

ABOVE OUTPUT IS FOR SCENE SEGMENT:
SCENE ID: 2282-2125400   STRIP NO.: 2 OF 4
SCAN LINES: 1760 – 1810
PIXELS: 530 – 644 (WEST TO EAST)
(EACH INDIVIDUAL PIXEL OUTPUT)

Figure 6.2 Supervised histogram parallelepiped classification applied to the Darfield agricultural subscene recorded over the Central Canterbury Plains, New Zealand on 31 October 1975. The character codes are as follows: (1) bare ground, (3) Kopara wheat, (6) alfalfa (or lucerne) and (9) exotic forest.

6.3.2.5 *Relation of the thematic map to ground features*

A colour-coded transparency of the themes, as produced under § 6.3.2.3 above, can be very difficult to relate to ground features. This is particularly evident with few classes classified amidst featureless agricultural plains.

Also, such a colour-coded thematic transparency only depicts those pixels within a field, or other supposedly homogeneous class block, that have been classified. There is no graphical evidence of which pixels were not classified within the block. These data would aid in assessing the influences of stress, edge effects, farming practice, etc, on the classification.

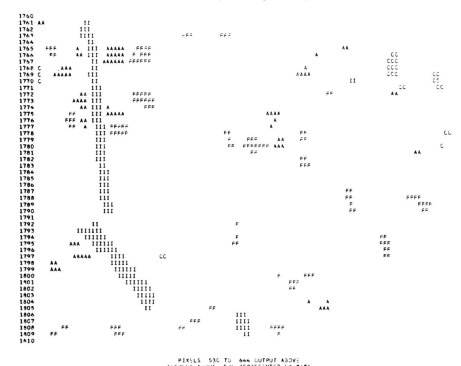

PIXELS 530 TO 644 OUTPUT ABOVE
(LEVELS ABOVE 530 REPRESENTED BY "●●")

Figure 6.3 The same region as in figure 6.2 has here been classified by the supervised histogram parallelepiped method with identical spectral signatures to those used for figure 6.2. Spatial postprocessing with a minimum proximity function discrimination level of 12×10^{-4} m^{-2} has then been applied (see Thomas 1980). In this case the character codes are: (A) bare ground, (C) Kopara wheat, (F) alfalfa (or lucerne), and (I) exotic forest.

A technique to cope with both these deficiencies is to output a modified MSS 5 file to tape. Such a single-band image can be modified so that all pixels that are classified are set to zero with all the remainder retaining their original MSS 5 values. As a result, when written out in the 'positive' transparency output mode the classified pixels are black with the MSS 5 remainder in grey shades.

Consequently, by producing such an image the effectiveness of the classification for supposedly homogeneous blocks may be assessed, and departures from this 'homogeneity' investigated.

The MSS 5 channel was selected as being most appropriate to vegetative land cover classification. As outlined in Chapter 2, a chlorophyll absorption band coincides with the MSS 5 channel. (It is usually an easy matter to modify the software to select another channel for another classification application.)

If the two image production processes, the colour-coded classification and the black and white modified MSS 5 data, are combined during the image creation onto the transparency an overlay image is produced. This effectively overlays the MSS 5 topographic data over the classified themes thus permitting the classified blocks to be related to topographic features.

6.4 The Superposition of Cartographic Information on a Thematic Map Product

The superposition of an overlay during the photographic printing process, from transparency to paper, further permits the relation of either classified areas (or the original satellite data) to topographic detail.

The typographic and cultural line work base is commonly taken from cartographic masters in negative transparency form and punch registered to the projected image. Then the photographic paper is exposed to the image followed by the exposure of the line overlay.

Overlays on *unrectified* Landsat imagery of New Zealand are 'helpful' to scales up to 1:500 000. With registered imagery, overlays have been combined to within topographic map accuracy standards up to 1:50 000.

The original technique was developed to support mapping in regions without many control points for position fixing, such as the Antarctic and the South Pacific.

6.5 The Derivation of Synthetic Channels

A Synthetic Channel is regarded as any channel that is produced from more than one of the basic bands, for example the Band Ratios and Principal Component Channels. These are the Synthetic Channels that are routinely available within image processing packages, e.g. the ERMAN package (IBM 1976). Additional difference–additive and ratio channels can be easily created to suit particular user needs within most software suites. Such extra channels could form the basis for different algorithms (see for example *The Tasseled Cap Transformation, Perpendicular Vegetative Index* and *Transformed Vegetation Index* discussed by Heydorn *et al* (1978)).

6.5.1 Band Ratios

Band Ratios are commonly regarded as being of most use in reducing the effects of topography on multichannel remote sensing data of ground cover (Holben and Justice 1979). Another major use is the elimination of some atmospheric effects from other multichannel data bases. Band Ratios can also be regarded as being of use in increasing the separation between ground cover classes—this will be supported by the discussion on Divergence in Chapter 7.

The simplest form of Band Ratio is of the format

$$R = I_1/I_2.$$

This has two immediate limitations: when $I_2 = 0$ and when I_1 is less than I_2. In the first instance R approaches infinity, if permitted and in the second, the results are compressed into the range of R between 0 and 1.

The Band Ratios used within ERMAN are presented in table 6.1 (after Winters 1979). Here the denominator is prevented from equalling zero and a broader distribution of the ratioed values is permitted by the use of multiplying constants. (The variation in these multiplication constants is brought about by the ranges for the Landsat 1, 2, 3 MSS channels 4, 5, 6 being 0–127 and that for MSS 7 being 0–63.)

Table 6.1 ERMAN Band Ratios accessed on the Sydney, Australia, system (from discussions with Winters (1979), Beach (1979)).

Ratio of MSS bands	ERMAN algorithm: $I(x_j)$ = CCT level radiance of that MSS (x_j) band
4/5	$32 \times I_4/(I_5 + 1)$
4/6	$32 \times I_4/(I_6 + 1)$
4/7	$16 \times I_4/(I_7 + 1)$
5/4	$32 \times I_5/(I_4 + 1)$
5/6	$32 \times I_5/(I_6 + 1)$
5/7	$16 \times I_5/(I_7 + 1)$
6/4	$32 \times I_6/(I_4 + 1)$
6/5	$32 \times I_6/(I_5 + 1)$
6/7	$16 \times I_6/(I_7 + 1)$
7/4	$64 \times I_7/(I_4 + 1)$
7/5	$64 \times I_7/(I_5 + 1)$
7/6	$64 \times I_7/(I_6 + 1)$

Another approach is to use a function such as outlined by Hord (1982):

$$R = C \tan^{-1}(I_1/I_2).$$

Here R ranges from 0 to $1.57 \times C$ and the multiplying constant C may be adjusted to suit the characteristics of the output device. For an 8-bit device, C could be chosen as 162.338 giving a range for R of 0–255. For a 6-bit device C could be chosen as 40.107 allowing R to range from 0 to 63 as required.

6.5.2 Principal Components

The other set of Synthetic Channels available routinely within ERMAN and similar software packages is that of the Principal Component channels.

The main aim usually in invoking the Principal Component channels is to reduce the dimensionality of the original data. The Landsat MSS is a four-channel instrument; thus, following Swain and Davis (1978), most of the data content could be expected to be digested into the first two Principal Component channels (using their cited $n/2$ rule of thumb where n is the number of the original data channels).

Let us attempt to visualise this transformation. In figure 6.4(a) the original data distribution for a series of classes is plotted as a bounding contour. The intercepts, such as are used in the parallelepiped classifier, are indicated for the two channels as $(a_1\ a_2)$ and $(b_1 b_2)$. If we could align a new set of axes along the major axis of the bounding ellipse (in this case) then the projection of the data variation on this new axis should be greater than either of the projections $(a_1 a_2)$ or $(b_1 b_2)$. That is, the separability between minor classes within the bounding contour (within the limits of the digitising system) should be greater and hence the classes should be more easily discriminated. Following an evaluation of the relevant channel data variances, a decision could be made to reduce the effective dimensionality of the data.

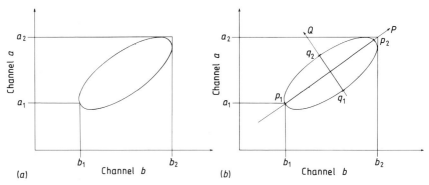

Figure 6.4(a) Plot in two-channel (a,b) space of the bounding contour to the data distribution sensed by detector channels $(a,\ b)$. (b) The superposition of the two Principal Component channels P and Q upon the basic data channels $(a,\ b)$.

In figure 6.4(b) this transformation is illustrated. A new set of orthogonal axes has been impressed upon the data distribution with its origin at the centre of gravity of the original data and one of the axes aligned along the major axis of the bounding ellipse. These are called the Principal Component axes and the transformation involves a rotation and a translation. (For some applications, such as image processing of remote sensing data, the translation is not fully implemented and may involve only an effective movement to the lower left end of the original bounding contour to avoid, technically, negative values.) We have here represented this data distribution as an ellipse for simplicity. Most real data distributions may be represented more accurately by elliptical bounding contours, or a series of them, than by circular bounding contours.

By invoking such a transformation then, the data variation (p_1p_2) is usually found to be greater than either (a_1a_2) or (b_1b_2). This will be illustrated with reference to data variance in the following derivation.

Let us now more rigorously consider the actual derivation of these Principal Component channels. Let us consider two channels (x_1, x_2) of the multispectral system and m data points

$$(x_{11}, x_{21}), (x_{12}, x_{22}), \ldots, (x_{1m}, x_{2m})$$

distributed within this two-dimensional feature space. We will probably find, unless the data points were highly uncorrelated in (x_1, x_2) space, that a boundary could be drawn around the data points which would have some tendency towards an elliptical shape, as we have suggested previously. As such, a major axis could be derived for this ellipse and a minor axis set at right angles to the major axis. These major and minor axes would then constitute a new set of coordinate axes (y_1, y_2) with the relation between the old and new coordinate axes being achieved by a linear combination of the data values from the old coordinate system:

$$y_1 = t_{11}x_1 + t_{12}x_2 \tag{6.1}$$

$$y_2 = t_{21}x_1 + t_{22}x_2.$$

Equation (6.1) implies only a rotation of the (x_1, x_2) coordinate axes into the new axes (y_1, y_2). Consequently the origins of both the (x_1, x_2) and (y_1, y_2) axes remain coincident. For the new (y_1, y_2) axes to more effectively represent the data distribution, the origin of the new axes should be translated to the centroid or 'centre of gravity' of the data $(\bar{x}_1, \bar{x}_2)$. This would remove the present implicit mathematical constraint that the major axis y_1 pass through the origin of the old (x_1, x_2) axes. Further, the minor axis, whilst still perpendicular to the major axis of the bounding ellipse, is translated to the data centroid, between the two foci, rather than being at the old origin, near the bounding periphery. This is accomplished by including translation terms in equation (6.1):

$$y_1 = t_{11}(x_1 - \bar{x}_1) + t_{12}(x_2 - \bar{x}_2) \tag{6.1a}$$

$$y_2 = t_{21}(x_1 - \bar{x}_1) + t_{22}(x_2 - \bar{x}_2).$$

However, to permit the reasons for using this Principal Components transformation to be seen more clearly, the inclusion of the translation terms is here deferred to a latter stage.

Returning to equation (6.1), if y_1 represented the major axis of the above elliptical boundary then the length of the major axis, or dynamic range of the data points along that new coordinate, would then exceed the dynamic range along either the x_1, x_2 or y_2 axes. Consequently, depending upon the ratio between the major and minor axes, most of the information content in the original data distribution in (x_1, x_2) two-dimensional space could be adequately represented by the transformed data in y_1 space. The dimensionality of the data can thus be reduced to what Swain and Davis (1978) term the 'intrinsic dimensionality' for the data system. Similarly, as the data dynamic range in y_1 would exceed that in x_1 or x_2 the separability of individual data points is increased in y_1 space. This implies that the use of this Principal Component transformation could lead to greater separability between two ground cover classes along the major axis of this ellipse if the centres of gravity of the two classes lie along or near this axis. (Swain and Davis (1978) found that the intrinsic dimensionality of Landsat 1, 2 four-band MSS data was approximately 2 and this was confirmed by discussions with Wheeler (1980).)

If data obtained from more than two channels are considered, as in the Landsat four-band MSS case, then

'new coordinate directions are defined, one after another, each time taking the direction of orientation of the new coordinate to be perpendicular to all the previously selected ones and in the direction of the remaining maximum data dynamic range' (Swain and Davis 1978).

The inclusion of the four Principal Component channels, derived from the basic four MSS bands, within the classification process could be expected to aid the classification process. This could be achieved through increased separability of the centres of gravity of the data for the ground cover classes brought about by the transformation. This would be particularly evident if a parallelepiped classifier was being used.

A simplified Principal Component transformation equation will now be set up from a four-band base dataset—still leaving the data centroid translation terms to a later stage—to illustrate the concept.

Equation (6.1) may be expressed in matrix notation as

$$\begin{bmatrix} y_1 \\ y_2 \end{bmatrix} = \begin{bmatrix} t_{11} & t_{12} \\ t_{21} & t_{22} \end{bmatrix} \begin{bmatrix} x_1 \\ x_2 \end{bmatrix} \tag{6.2}$$

or equivalently

$$\mathbf{Y} = \mathbf{TX} \tag{6.3}$$

where **Y**, **T**, **X** denote the matrices whose elements, i.e. t_{ij}, refer to the ith row and jth column (Aitken 1951).

Extending this matrix notation to n dimensions results in the matrix of the basic (old) data **X** being a column vector of n elements, that is a matrix of order $n \times 1$. Similarly the transformed (new) data matrix **Y** must also be of order $n \times 1$. Necessarily then, the transformation matrix **T** must be a square matrix of order $n \times n$.

As a pure rotation (i.e. no translation) is being performed by **T** the origin of the basic and the transformed coordinates is the same point in feature space. Consequently if the axes of the transformed coordinates are orthogonal then the Euclidean distance between the origin and the data points is unchanged by the transformation. For this to be true **T** must be an 'orthogonal' matrix such that $\mathbf{TT'} = \mathbf{I}$ where **T'** is the transpose of **T** ($t'_{ij} = t_{ji}$) and **I** is the unit matrix. Consequently, to effect the required Principal Component transformation, we seek an n-dimensional square orthogonal matrix.

Any transformation depends upon the interrelationship between the basic coordinate axes. If these relations are broken down to pairs of coordinate axes the relationship between data points within the plane defined by these axes may be expressed as a single number which is a symmetrical function of the basic data points. This quantity is known as the covariance s_{12} and is defined by

$$s_{12} = \frac{1}{m-1}\sum_{i=1}^{m}(x_{1i} - \bar{x}_1)(x_{2i} - \bar{x}_2) \qquad (6.4)$$

where 1 and 2 are the coordinate axes and the total number of data points is m (Hald 1962). These s_{12} elements can constitute a variance/covariance matrix **S** (Brandt 1970). The diagonal elements of this matrix, s_{jj}, are the variances with the standard deviations for the data recorded in that channel, s'_j being the positive square roots of the variances (Brandt 1970):

$$s'_j = +\sqrt{s_{jj}}. \qquad (6.5)$$

This variance/covariance matrix **S** is then the basis for the Principal Component (PC) transformation between coordinate systems.

Note that the variance/covariance matrix, used to set up the Principal Component transformation for a specified section of an image, is computed for that complete section treating it as one 'class' (in the terminology of Chapter 7). Also, henceforth in this chapter the PC variance/covariance matrix is understood to refer to one matrix over all nominated channels.

Ready and Wintz (1973) give the PC transformation relation as

$$\mathbf{Y} = \mathbf{T}\,(\mathbf{X} - \mathbf{U})$$

where $\mathbf{X}$ is the vector of data points, in the basic coordinate system; $\mathbf{U}$ is the mean vector, in the basic coordinate system; $\mathbf{Y}$ is the vector of Principal Components—the new coordinate system; and **T** is the $n \times n$ orthogonal matrix that effects the transformation. **T** is derived from the variance/covariance matrix **S** such that the rows of **T** are the normalised (length normalised to unity) eigenvectors of **S**.

The use of the transformation defined by equation (6.6) permits the origin of the new coordinate system to be translated to the 'centre of gravity' of the original data points, in feature space. This translation is effected by $(X-U)$ with the rotation being performed by **T**—as discussed previously. The combined transformation will lead to negative coordinate values for some data points in the new PC system. Compensation must be made for such negative values when generating images for pictorial display—as suggested earlier.

The major task in establishing the transformation is setting up the rotation matrix **T** from **S**. This derivation is now followed through on a step-by-step basis.

The eigenvectors G_i of the matrix **S** are defined such that equation (6.7) is satisfied where c_i is defined to be an eigenvalue of the matrix **S**:

$$\mathbf{S}G_i = c_i G_i. \tag{6.7}$$

On inserting the unit matrix **I**, equation (6.7) becomes

$$(\mathbf{S}-c_i\,\mathbf{I})G_i = 0 \tag{6.8}$$

from which the set of eigenvalues may be derived, via equation (6.9), if the eigenvector G_i is non-zero:

$$|\mathbf{S}-c_i\,\mathbf{I}| = 0. \tag{6.9}$$

This eigenvalue equation (6.9) yields a set of values for c_i. Each of these will yield a corresponding eigenvector G_i from equation (6.8). The set of n homogeneous linear equations that are produced from equation (6.8) will be of the form

$$\mathbf{A}G_i = 0 \tag{6.10}$$

where

$$\mathbf{A} = (\mathbf{S}-c_i\,\mathbf{I}).$$

Such a set of homogeneous equations, with $|\mathbf{A}| = 0$, never has a unique solution as **A** is a singular matrix. If G'_{ij} is one solution for the jth variable of the ith eigenvector (where $i, j = 1, 2. \ldots, n$, to n dimensions) then gG'_{ij} is also a solution for any value of the constant g. The solution is therefore made unique by adding an extra non-homogeneous restriction. A common technique followed here is to fix the value of one variable as unity (after Hall 1963). When an eigenvalue is zero the above solution may still not ensure a unique solution. Recourse is then made to the condition that the scalar, or dot, product of the eigenvectors, $G_i \cdot G_j$, where $i \neq j$, must be zero—from $\mathbf{TT}' = \mathbf{I}$. If more than one eigenvalue is zero the condition $\mathbf{TT}' = \mathbf{I}$ cannot be satisfied. This leads us to the conclusion that a number of data points at least equal to the number of dimensions must be used in deriving **T**. This may also be thought of as: one data point lies on a one-dimensional line that includes the origin of the transformed data, a minimum of two points together with the origin are needed to define a PC plane, at least three data points are required to define a three-dimensional PC cube, and so on.

Equation (6.10) then gives an eigenvector set G_i' with components G_{ij}'. The vector is then length normalised to one to give the normalised eigenvector G_i that relates to the eigenvalue c_i. The elements G_{ij} then constitute the ith row of the transformation matrix $\mathbf{T}$.

The eigenvalue matrix $\mathbf{C}$ has non-zero elements

$$c_{ii} = c_i. \tag{6.11}$$

If the eigenvalues c_i are ranked in numerically decreasing order they represent the variances c_{ii} of the Principal Components in the transformed coordinate system. Consequently the Principal Components may be regarded as being uncorrelated with each component having a variance less than the previous component. The information content proportion of the various transformed components may thus be assessed from the data variances, as a percentage of the total.

The transformation

$$\mathbf{TST}' = \mathbf{C} \tag{6.12}$$

also generates the eigenvalue matrix $\mathbf{C}$.

Let us illustrate the concept of Principal Components with an example. If we take two original data points in (x_1, x_2) space at $(3,1)$ and $(6,2)$ and seek the transformation that positions the point in Principal Component (y_1, y_2) space then the new sets of coordinates should be $(-1.581, 0.000)$ and $(+1.581, 0.000)$—in Principal Component coordinates where the origin of the PC coordinates has been translated to $(4.50, 1.50)$ in the original coordinates. These values were obtained from basic Pythagorean geometry.

Now let us follow through the Principal Component derivation which builds on the variance/covariance matrix $\mathbf{S}$ whose elements have been derived from equation (6.4):

$$\mathbf{S} = \begin{bmatrix} 4.5 & 1.5 \\ 1.5 & 0.5 \end{bmatrix}. \tag{6.13}$$

The eigenvalues (c_1 and c_2) are then obtained from equation (6.9):

$$|\mathbf{S} - c\,\mathbf{I}| = \begin{vmatrix} 4.5-c & 1.5 \\ 1.5 & 0.5-c \end{vmatrix} = 0.$$

Hence $c_1 = 5.0$, $c_2 = 0.0$.

Note also that the trace of $\mathbf{S}$, the sum of the diagonal elements, is equal to the sum of the eigenvalues—as it should be. This means, in this application, that the total data variance is preserved by the transformation.

Two sets of homogeneous equations are produced from equation (6.8), one for each eigenvalue:

$$-0.5G'_{11} + 1.5G'_{12} = 0$$
$$1.5G'_{11} - 4.5G'_{12} = 0 \qquad \text{for } c_1 = 5.0 \qquad (6.14)$$

and

$$4.5G'_{21} + 1.5G'_{22} = 0$$
$$1.5G'_{21} + 0.5G'_{22} = 0. \qquad \text{for } c_2 = 0.0 \qquad (6.15)$$

Solving equations (6.14) and (6.15) independently for $G'_{11}, G'_{12}, G'_{21}, G'_{22}$ and setting one variable to 1.0 in each case yields, upon normalising the length of the eigenvectors to unity,

$$G_{11} = 3/\sqrt{10} \quad G_{12} = 1/\sqrt{10} \quad G_{21} = 1/\sqrt{10} \quad G_{22} = -3/\sqrt{10}. \qquad (6.16)$$

For **T** we have

$$\mathbf{T} = \begin{bmatrix} G_{11} & G_{12} \\ G_{21} & G_{22} \end{bmatrix}. \qquad (6.17)$$

T is orthogonal since **TT′** = **I**—as required. It is also quickly verifiable that **TST′** = **C**.

Applying equation (6.6) to the point (3,1) we have

$$\begin{bmatrix} y_1 \\ y_2 \end{bmatrix} = \begin{bmatrix} 3/\sqrt{10} & 1/\sqrt{10} \\ 1/\sqrt{10} & -3/\sqrt{10} \end{bmatrix} \begin{bmatrix} (3-4.5) \\ (1-1.5) \end{bmatrix} = \begin{bmatrix} -1.581 \\ 0.000 \end{bmatrix}.$$

Similarly the point (6,2) transforms to the point (+1.581, 0.000) in PC space, with both results being in agreement with the earlier calculation via Pythagorean geometry.

A comparison of the dynamic ranges (DR) for the data in the basic and transformed coordinate systems suggests that the inclusion of the first Principal Component PC 1, or y_1, in the classification process could be advantageous:

$$\text{DR}(x_1) = 3.0 \qquad \text{DR}(x_2) = 1.0 \qquad \text{DR}(y_1) = 3.162 \qquad \text{DR}(y_2) = 0.000.$$

6.6 Derivation of a Sample Set of Synthetic Channels

The aim in this context of creating the synthetic channels is to use them in the classification process to effect greater separation of the classes than would be possible by using the basic data bands alone. This section presents a sample set of synthetic channels obtained from the basic Landsat bands.

Ching (1981) determined the spectral signature means and standard deviations for four classes of forest from training fields within Landsat image 2389–21172 recorded over the central North Island of New Zealand on 15 February 1976 (GMT). (A maximum of four classes were selected to satisfy the condition for the minimum number of data points required for a Principal Component analysis from the basic four Landsat bands—see earlier.) These four classes were *Nothofagus fusca/menziesii* (red/silver beech), *Pinus contorta, Beilschmiedia tawa* (tawa), and *Pinus radiata* (1962 planting); these are referred to as classes N, P, B and R respectively.

Table 6.2 presents the basic data means for each of the four classes. The means were derived from the Band Ratio algorithms of table 6.1, and the transformed Principal Component means. These latter means were obtained from the basic data means, using the foregoing mathematics, as outlined in the appendix to this chapter.

Table 6.2 The means, and derived means, for the basic four MSS bands (MSS x), Band Ratios ('U/V') and the Principal Components—all synthesised from the Basic Bands (PC y). The Band Ratios are derived using the algorithms presented in table 6.1. The Principal Components have been derived from the variance/covariance matrix—see the appendix to this chapter.

Channel No	Channel description	Class means			
		Class N	Class P	Class B	Class R
1	MSS 4	9.562	9.346	9.920	9.580
2	MSS 5	8.325	7.132	8.749	8.112
3	MSS 6	24.979	27.112	30.297	31.879
4	MSS 7	13.100	14.379	16.349	18.201
5	'4/5'	32.813	36.777	32.561	33.644
6	'4/6'	11.778	10.639	10.143	9.324
7	'4/7'	10.851	9.723	9.149	7.983
8	'5/4'	25.223	22.059	25.638	24.535
9	'5/6'	10.254	8.118	8.946	7.895
10	'5/7'	9.447	7.420	8.069	6.760
11	'6/4'	75.680	83.857	88.782	96.420
12	'6/5'	85.719	106.688	99.447	111.954
13	'6/7'	28.345	28.207	27.941	26.564
14	'7/4'	79.379	88.948	95.818	110.101
15	'7/5'	89.909	113.165	107.328	127.838
16	'7/6'	32.272	32.735	33.432	35.429
17	PC 1	−4.300	−1.891	+1.935	+4.256
18	PC 2	−0.443	+0.866	−0.659	+0.237
19	PC 3	+0.196	−0.172	−0.269	+0.245
20	PC 4	+0.000	−0.000	−0.000	+0.000

Within ERMAN (IBM 1976) the Principal Components may be derived from either the variance/covariance matrix or the correlation matrix. The latter is easily derived from the former by dividing the covariance term by the products of the square roots of the variances in the constituent channels. The discussion here is therefore confined to the source matrix; the variance/covariance matrix.

6.7 References

Aitken A C 1951 *Determinants and Matrices* (Edinburgh: Oliver and Boyd)

Beach D W 1979 Informal Communication (Sydney: IBM)

Bernstein R 1978 *Digital Image Processing for Remote Sensing* (New York: IEEE)

Brandt S 1970 *Statistical and Computational Methods in Data Analysis* (Amsterdam: North-Holland)

Ching N P 1981 Informal Communication (Wellington: New Zealand Forest Service)

Ellis P J, Thomas I L and McDonnell M J 1978 *Landsat II over New Zealand: monitoring our resources from space* DSIR Bull. 221 (Wellington, NZ:DSIR)

Fahnestock J D and Schowengerdt R A 1983 Spatially variant contrast enhancement using local range modification *Opt. Eng.* **22** 378

Hald A 1962 *Statistical Theory with Engineering Applications* (New York: Wiley)

Hall G G 1963 *Matrices and Tensors* (Oxford: Pergamon)

Hall R E and Awtrey J D 1980 Real-time image enhancement using 3×3 pixel neighbourhood operator functions *Opt. Eng.* **19** 421

Heydorn R P, Trichel M C and Erickson J D 1979 Methods for segment wheat area estimation *Proc. Tech. Sessions LACIE Symp.* vol II, Publ. No JSC-16015 (Houston: NASA) p 621

Holben B and Justice C 1979 *An Examination of Spectral Band Ratioing to Reduce the Topographic Effect on Remotely Sensed Data* NASA Tech. Mem. 80640 (Goddard Space Flight Center, MD: NASA)

Hord R M 1982 *Digital Image Processing of Remotely Sensed Data* (New York: Academic)

Horn B K P and Woodham R J 1979 Destriping Landsat MSS images by histogram modification *Comp. Graphics Image Process.* **10** 69

IBM 1976 *Earth Resources Management II (ERMAN II) User's Guide* Program No 5790-ARB Manual No SB11-5008-0 (Brussels: IBM)

Ready P J and Wintz P A 1973 Information extraction, SNR improvement, and data compression in multispectral imagery *IEEE Trans. Commun.* **COM-21** 1123

Reeves R G 1975 *Manual of Remote Sensing* (Falls Church, VA: American Society of Photogrammetry) p 2144

Swain P H and Davis S M 1978 (ed.) *Remote Sensing: The Quantitative Approach* (New York: McGraw-Hill)

Thomas I L 1979 Cartography from Landsat: introducing the DSIR computerised land use mapping package *Proc. 49th ANZAAS Congress — Geographical Sciences (Auckland, NZ)* p 276

——1980 Spatial postprocessing of spectrally classified Landsat data *Photogramm. Eng. Rem. Sens.* **46** 1201

Thomas I L, Barrett R C and Benning V M 1979 Classification and textural enhancement using the DSIR Landsat ANalysis SYStem (LANSYS 1) computer package *Proc. Landsat '79 Conf. (Sydney)*

Thomas V L 1973 *Generation and physical characteristics of the ERTS MSS system corrected Computer Compatible Tapes* Doc. No X-563-73-206 (Goddard Space Flight Center, MD: NASA)

Timmins S M 1981 *MSc Thesis* Univ of Waikato, NZ

USGS 1979 *Landsat Data Users Handbook* (Arlington, VA: US Geological Survey)

Wheeler S G 1980 Informal communication, IBM, Houston

Winters B J 1979 Informal communication, IBM, Sydney

Appendix

Derivation of an example set of Principal Components

The means from the basic data for the four classes N, P, B and R were transformed into their Principal Component means in the following manner. The class N is *Nothofagus fusca/menziesii*, P is *Pinus contorta*, B is *Beilschmiedia tawa* and R is *Pinus radiata* (1962 planting) (see earlier).

The means within each of the basic MSS data bands for each class were obtained by Ching (1981) using the Thomas (1979) analysis system. These means together with the average of the means are listed in table 6.A.1.

Table 6.A.1 Basic MSS data means and averages of these means for the four forest classes being considered.

MSS channel	Class				Average
	N	P	B	R	
4	9.562	9.346	9.920	9.580	9.602 000
5	8.325	7.132	8.749	8.112	8.079 500
6	24.979	27.112	30.297	31.879	28.566 750
7	13.100	14.379	16.349	18.201	15.507 250

(Throughout this derivation the calculations were performed on a Hewlett-Packard 33E calculator to eight decimal places—to minimise any rounding errors.)

These basic data points led to the variance/covariance matrix, table 6.A.2, through the use of relation (6.4).

Table 6.A.2 Covariance matrix, **M**, for the data from table 6.A.1 (the elements M_{ij} = M_{ji}).

+0.056 248 00	+0.148 308 67	+0.331 092 00	+0.197 845 33
	+0.469 104 33	+0.587 877 83	+0.376 378 50
		+9.651 004 25	+6.885 581 42
			+5.010 877 58

The solutions of the eigenvalue equation

$$|\mathbf{M} - c_i\,\mathbf{I}| = 0 \tag{6.9}$$

were a set of eigenvalues:

$$c_1 = 14.641\,683\,23 \qquad c_2 = 0.478\,693\,74$$
$$c_3 = 0.066\,857\,19 \qquad c_4 = 0.000\,000\,00.$$

The sum of the diagonal elements of the variance/covariance matrix **M** is equal to the sum of the eigenvalues ($= 15.187\,234\,16$). Consequently the total data variance is preserved since the diagonal elements of **M** are the variances of the original data bands and the eigenvalues are the variances of the transformed data channels.

It is interesting to compare the data variance proportion carried by each of the basic and transformed data channels. This is compared and expressed as a percentage of the total in table 6.A.3. Remembering that this example is drawn from a forestry application, the emphasis on MSS 6 in the original data is interesting, particularly since the usual colour composite concentrates on MSS 4, 5, 7 and the SPOT spacecraft also tends to exclude this spectral region.

Table 6.A.3 Comparison of the relative proportional percentages of the total data variance carried by the original data channels and the transformed channels.

Original data channels		Transformed data channels	
Variance component	%	Variance component	%
M_{11}	0.3704	c_1	96.4078
M_{22}	3.0888	c_2	3.1519
M_{33}	63.5468	c_3	0.4402
M_{44}	32.9940	c_4	0.0000

The matrix equation (6.8) now yields a set of eigenvectors, normalised to unit length, one ($\mathbf{G}_i$) for each of the previously determined eigenvalues (c_i):

$$(\mathbf{M} - c_i\,\mathbf{I})\,\mathbf{G}_i = 0. \tag{6.8}$$

In solving the linear homogeneous equations, the fourth variable in each G_i was set to 1—to exercise a uniqueness constraint (see main text). Further, for the equation set derived for $c_4 = 0.0$, it was necessary to invoke the further uniqueness constraint that $G_i \cdot G_j = 0$, where $i \neq j$ (see earlier comment). (G_i was length normalised to unity.)

Each G_i was then regarded as a row of the transformation matrix T and this matrix is presented in table 6.A.4.

Table 6.A.4 The transformation matrix T that effects the rotation for the Principal Component transformation for the four classes considered here. The appropriate eigenvalue for each row eigenvector is indicated.

+0.026 811 29	+0.049 386 00	+0.811 015 53	+0.582 319 49	c_1
−0.291 588 17	−0.948 805 27	−0.018 727 93	+0.119 975 69	c_2
−0.274 089 91	+0.191 757 25	−0.551 366 68	+0.764 263 48	c_3
+0.916 041 68	−0.246 086 58	−0.194 673 73	+0.249 822 29	c_4

The products

$$T\,T' = I \qquad T\,M\,T' = c_i\,I$$

were checked and agreement to at least $\pm\ 1.0 \times 10^{-7}$ was obtained.

The transformation of the original data points, through equation (6.6), was now performed:

$$Y = T(X - U). \tag{6.6}$$

The rotation matrix operated on the translated vector for each class and led to the revised Prinicipal Component means presented in table 6.A.5.

Table 6.A.5 Transformed Principal Component means derived from the class means of table 6.A.1.

	Class			
PC	N	P	B	R
1	−4.300 457 75	−1.890 483 73	+1.935 016 97	+4.255 924 51
2	−0.442 888 52	+0.865 521 45	−0.659 364 63	+0.236 731 70
3	+0.196 432 55	−0.171 702 57	−0.269 462 53	+0.244 732 55
4	+0.000 000 04	−0.000 000 03	−0.000 000 02	+0.000 000 00

The variance of these transformed data, in each channel, was checked to be equal to the eigenvalue for that channel—within $\pm\ 1.0 \times 10^{-7}$.

The values from table 6.A.5 became the Principal Component means for these classes in the synthetic channel study and were presented in the main body of the text as table 6.2.

7 The Classification of Ground Cover Classes and Measures of their Separability

7.1 Classification of Ground Cover Classes

The classification of remotely sensed data reduces the number of data channels to a single channel with the data grouped into blocks of user defined themes. These themes may be major land cover types (e.g. cereal crops) or in more refined groups of subclasses (e.g. 'cereal crop. wheat crop. Kopara wheat. moisture level in leaf. percentage deficiency in nutrient X. grown on soil type Y. by farmer Z'—where the points indicate successive subdivisions of the categories).

Themes are user defined and serve the (administrative) purposes of that user.

'Each land (cover) classification is made to suit the needs of the user, and few users will be satisfied with an inventory that does not meet most of their needs. . . In almost any classification process, it is rare to find the clearly defined classes that we would like. In determining land cover, it would seem simple to draw the line between land and water, until one considers the problems of seasonally wet areas, tidal flats, or marshes with various kinds of plant cover. Most types of land (cover) appear in a continuous gradation from zero, or no (cover), to 100%, or full (cover). The problem to be solved is where to place the boundary around the (cover) we wish to classify. How small an area of land with discarded equipment on it should be called a junkyard? How do we resolve the problem of heterogeneous mixtures of equally significant land (covers)? Decisions that may seem arbitrary at times must be made, but if the descriptions of categories are complete and guidelines are explained, the inventory process can be repeated. The classification system must allow for the classification of all parts of the area under study and should also provide a (descriptive reference for each type of land cover)'—Anderson *et al* (1972).

7.1.1 Considerations in the classification process

In the classification process:

(i) The user must decide how much heterogeneity can be permitted in each of the physical sampling units, for his/her defined classes, whilst maintaining the homogeneity of that class over an acceptable area for the classification process. In this decision the concept of the levels of classification (Chapter 3) can assist.

(ii) The user must outline representative examples of the required types of land cover to the analysis system.

(iii) The system must accept these defined areas, and hence representative statistics, and thereafter present repeatable results.

(iv) The classification process must have only the one input and this is the user defined and recorded land cover type entered as a training field. These are the only data explicitly entered on a field-by-field basis by the user. Other data planes such as meteorological data, soil data etc would have been entered previously on a regional basis.

(v) On the ground each resolution element probably contains slightly different mixtures of ground cover types. The class defined to the system must therefore be a little less sensitive to these variations. The classification process is thus presented with data that, away from the training areas, have tendencies towards, or away from, the specified homogeneous ground cover classes. The analyst must appreciate the significance of this fact in his/her work as it bears on the thresholding decisions that are discussed later.

(vi) The system should advise the user whether or not the specified ground cover types can, in fact, be resolved acceptably by the data.

(vii) The user must evaluate this assessment of species resolution and decide whether to accept that level of separation, modify the administrative decision as to the required ground cover types or respecify training fields.

(viii) The system must aggregate each pixel into one of the specified classes which most closely fits the characteristics of that pixel.

(ix) The system must be able to assess the likelihood of each pixel being correctly allocated to one of the specified classes. Those with high likelihoods can then be accepted and the remainder rejected from the class maps.

7.2 Introduction to Multichannel Classifiers

Remotely sensed data usually involve a pixel having its characteristics recorded over a number of spectral channels. Similarly a class, specified by the analyst, extends over the same multiple channels. The difference between the two is that a pixel is then specified as a point in n-dimensional spectral space whilst the class is specified as a 'blob' in the same n-dimensional space (see figure 7.1). The characteristics of the blob are determined by the distribution of statistics for that class in each channel. These distributions are the projections in each channel of the blob's characteristics.

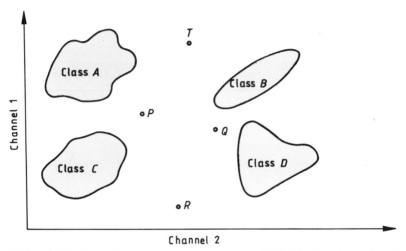

Figure 7.1 Relation between ground cover class 'blobs' in two-dimensional space—which are made up of the pixel spectral values used to specify the classes, and the pixels which are to be classified into one of the specified classes.

Let us now relate such a concept of class distributions in multichannel spectral space to a physical analogy, referring to figure 7.2 as a three-dimensional representation.

If we confine ourselves to a three-dimensional database, the class distributions within that spectral space may be likened to 'lumpy custard'. Each 'lump' within the three-dimensional data 'bowl' has a slightly different location—and hence may be regarded as a user specified class with different location parameters, or more correctly, different spectral signatures. Each lump has a higher density region around the centre and progressively less dense three-dimensional shells away from that centre. Somewhere between the lumps the less dense mixtures merge together. Any unit of the custard then has a finite probability of having the characteristics of one of the lumps and hence of belonging to that class.

In remote sensing terms, each 'lump' in the data 'custard' considered so far is a discrete ground cover type specified by the user. As such it is assumed that its statistical distribution rises clearly above the statistical continuum in each data channel. (In terms of figure 7.2 this would be above the 5% probability contour level.)

However in real life there are usually far more discrete class 'lumps' than are specified by the user. For example, for the basic four MSS bands of Landsat, if we make the assumption that each ground cover type 'lump' must extend over four CCT levels and, acknowledging that the original data were digitised to a resolution of six bits, then there can be as many as 65 536 possible 'lumps' in the four-dimensional data 'custard'. This figure of 65 536 was derived as follows: six-bit digitising means that the radiance recorded from the ground fell into one of 2^6 ($=64$) levels. As each ground cover 'lump' is arbitrarily allocated a four CCT level spread, with no spacing between the 'lumps', then 16 possible 'lumps' can be recorded in each channel. Over four channels a total of 16^4 ($=65\,536$) 'lumps' may thus be defined, under the above assumptions. If we made the same assumption of four CCT levels per 'lump' with 8-bit digitised eleven-channel aircraft scanner data there would be 7.4×10^{19} possible ground cover 'lumps' in the eleven-dimensional data 'custard'.

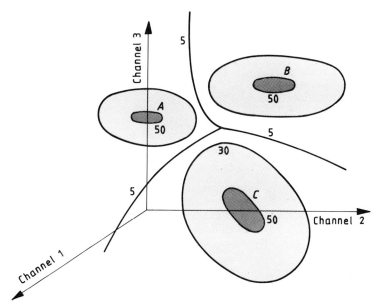

Figure 7.2 Three classes in three-dimensional feature space with their percentage probability contours outlined. The core regions ($>50\%$) constitute the 'lumps' in the data 'custard' that fills feature space. Notice the decision boundaries that must eventuate between classes. Further, mentally project yourself as a pixel into that data 'custard' at various locations and decide into which class (A,B,C) you would most likely be joined.

Obviously this discussion is highly stylised. On the ground there will be an even greater number of possible classes and certainly many more classes than the sensing system can differentiate. The point to note, however, from these examples is that the acquisition and computing systems are theoretically capable of separating more classes than are generally specified by the user in setting up the classification process.

The user confronted with such a data 'custard' selects only those 'lumps' that constitute the ground cover types that he, or she, requires. The system then processes the data 'custard' and assigns each pixel to one of the user specified ground cover types. If the pixel actually belongs to one of the other classes recognised by the system, and not specified by the user, it will have a low likelihood associated with it (see later) and should be 'thresholded out' of the final classification products.

'Thresholded out' refers to those pixels whose likelihoods are below a user nominated threshold level and hence have been excluded from the accepted classification results. (This is discussed in detail following the derivation of the likelihoods in this chapter.) Suffice it at this stage to remark that the individual classes have probability distributions associated with them in n-dimensional space. From the overlap of the class probability distributions at a given point for an unclassified pixel, a likelihood value may be derived for each pixel. This is derived according to certain user determined decision rules (to be discussed later) and is influenced by *all* the user defined class probability distributions. In short: a class has a probability distribution derived from its pixels' occurrence distribution and a classified pixel has a likelihood associated with it after classification, being derived from the class probability distributions.

If a user specifies classes A and B but not C in figure 7.2, a pixel then sitting in the high probability 'core' region of class C would have only low likelihoods of being in classes A or B. If A and B were the only classes specified by the user then a pixel having a resultant low likelihood of being in either of these classes should then be thresholded out of the final A/B classification. (In the original data distribution all three classes would have existed.)

Let us now replace the 'custard' analogy with the identical concept of overlapping probability distributions. Each pixel within a training field is specified as an occurrence in n-dimensional data space. The aggregation of these pixels (some having identical values in each of the n data channels) leads to an overall probability distribution for that class. Such a distribution could have its bounding contours defined by the curves of figures 7.1 or 7.2. However, each of these bounding contours also has a specific probability associated with it. Similarly each data point inside the bounding contour also has a probability associated with it. (These derived probabilities have been generated from the occurrence distribution of the original data pixels in the training field for that class.) We could therefore specify the class as an aggregation of data points each having an associated probability. Then when one of these data points (having specific values in each of the n data channels) coincided with the values of the pixel being evaluated, that pixel would take on the data point's probability of being allocated to that class. This technique can be implemented by storing all the data points and their associated probabilities in a matrix form or by more neatly specifying the bounding contours for different probability levels (e.g. 5, 10, 20, 40, 80% etc). These latter bounding contours could be specified as a series of termination points to straight lines that constitute the contours. Either of these approaches would be known as a non-parametric approach to classification. If the number of dimensions is small this approach is very effective in achieving rapid classification of data using minimal computer storage/time resources. However if the number of dimensions increases the storage constraints for a great number of classes rapidly suggests the use of a parametric approach.

The parametric approach relies on typifying the probability distribution by a representative function. In two-dimensional space, rather than storing the occurrence values of a histogram, we may just store the mean and standard deviation and use the normal distribution function to replicate the original probability distribution. Here the mean and standard deviation are the parameters. This approach allows easy extension to many dimensions and classes and is the approach concentrated on in this chapter.

The non-parametric approach to classification can be very effectively used as a preliminary step in refining ground cover training fields and optimum channels etc. Usually the most useful method is to record the lowest acceptable probability contours (e.g. 10%) in the two most effective channels for discriminating the major classes of interest. These contours are then bounded by parallelepiped limits (see Chapter 6) and are classification executed. The use of a spectral transformation (e.g. Principal Component) ahead of the parallelepiped classification will improve these classification results—as indicated in Chapter 6.

Returning then to the parametric approach to classification : the mean and the standard deviation are the parameters that are used to typify the probability distribution. The mean vector in multidimensional space locates the single peak of the unimodal probability distribution; and the standard deviation, for that ground cover type, specifies the region of influence of that probability distribution in each dimension.

We now attempt to visualise the situation that the parametric classifier must process.

Ideally the user will specify one or more of the unimodal distributions, through the choice of suitable training fields as required ground cover classes. If the specified ground cover class covers several unimodal 'lumps', e.g. Class D in figure 7.1, then a degraded result will be produced since the inserted statistical profile for that class will take on the composite characteristics of several other classes. The separation between the means of the inserted composite classes will be less than would have otherwise been possible. Also the probability distributions will be broader and less sharply defined. This will be revealed as a series of larger standard deviations in the 'class' statistics than would have been expected. Both factors will lead to an apparent decrease in class separability.

Having considered the concept of the distributions of class statistics in multidimensional spectral space, it is now appropriate to consider how such classes may be differentiated one from another by computing systems. A variety of classifiers are available and the more usual ones are now described.

7.2.1 The parallelepiped classifier

The parallelepiped classifier, presented in Chapter 6, is the simplest classifier and uses the least computer resources of all classifiers. The user specifies the mean ± acceptable limits to the class statistical distribution and these constitute the parallelepiped boundaries to the class 'lumps'. A disadvantage is that those pixels having the characteristics of the boundary region between classes may or may not be members of that class—there is no weighting for the certainty level of classification for each pixel. Some pixels can therefore be composites of the specified classes. This possibility is not permitted by the parallelepiped classifier.

7.2.2 The Euclidean distance classifier

A better classifier is that based on the Euclidean distance. This assesses the distance, in multidimensional space, between the position of the data pixel and the location of the statistical mean for each specified class. The pixel is then allocated to the 'closest' class. This classifier makes no allowance for the distribution, in multidimensional spectral space, of the data points that led to the class probability distribution. That is, a well-defined (low standard deviation) class may be located very close in feature space to the pixel. If there is a less pure class (large standard deviation) located further away, a classifier based on statistical concepts would be in a quandary. The probability distribution of the ill-defined class will impose a greater likelihood value on the pixel than that of the well-defined class but the well-defined class is closer. Either class selection could be correct but the Euclidean distance technique will select the ('closer') well-defined class.

7.2.3 Limitations on parallelepiped and Euclidean classifiers

The limitation on both the parallelepiped and Euclidean distance classifiers is that they operate separately on independent discrete single-dimensional representations of the probability distribution and then bring the answer together in multidimensional space. For example, the Euclidean distance technique evaluates $|x_{1k} - \bar{x}_{1i}|$ (where x_{1k} is the data point in the first-dimensional space and $\bar{x}_{1i}$ is the mean of the distribution for class i in that space). These are all then squared and summed and the square root taken to give the distance x_k to $\bar{x}_i$ in n-dimensional space.

Spectral signatures represent the vegetation characteristics over the whole spectrum. This spectral domain is then sampled four times by Landsat to give the basic four data channels and eleven times by the aircraft scanner cited here (Chapter 5). As such the numbers in the data channels are not entirely independent but are related through the spectral signature they represent. Consequently a technique that acknowledges this correlation will utilise a probability distribution that takes account of such a basic correlation rather than one that considers the channels independently.

7.2.4 The Mahalanobis distance classifier

One such technique to acknowledge the interrelation of the spectral channels is that based on the Mahalanobis distance. This classification technique, for each specified class, firstly evaluates the distance between a point in spectral or feature space, and the mean position for that class. In this respect it is the same as the Euclidean distance approach. However, it then 'modifies' this distance by dividing by the variance of the class *in the appropriate n-dimensional (point-to-mean) direction*. Since such a variance is derived from the variance/covariance matrix operating on the direction vector, the Mahalanobis distance technique thus measures the remoteness of the point in terms of the distribution characteristics of the class training field data in n-dimensional space.

Formally, the Mahalanobis distance M^2 is equal to the square of the distance expressed in units of the variance for that class.

Whereas the square of the Euclidean distance is given by

$$(X_k - U_i)^{\mathrm{T}}(X_k - U_i)$$

the Mahalanobis distance M^2 is given by

$$M^2 = (X_k - U_i)^{\mathrm{T}} V_i^{-1}(X_k - U_i)$$

where X_k is the vector representing the point k; U_i is the mean vector for class i; V_i is the variance/covariance matrix for class i.

As will be seen shortly, M^2 appears as a *component* of the full expression for the probability function used by the Maximum Likelihood classifier (see equation (7.4)).

7.2.5 Maximum Likelihood classifier

The Maximum Likelihood classifier

(i) prepares an n-dimensional probability function that takes account of the correlation terms between data channels. It therefore reflects the correlation seen in the actual spectral signature;

(ii) evaluates the n-dimensional probability function at the location of the data pixel and then calculates a 'likelihood' value, for that pixel, of classification into the most likely of the specified classes.

It uses a parametric statistical approach to prepare the probability density distribution functions for each individual class. It then employs the Bayes optimal strategy to maximise the likelihood of correct classification when allocating every pixel within the dataset to one of the user specified classes. The user then decides upon what is an acceptable level of the likelihood for each class and permits only those pixels above this level to be accepted into a final classification product.

It is always assumed here that prior to invoking the classifier the user has specified, to the system, adequate homogeneous training fields that permit an n-dimensional unimodal probability function to be allocated to that class.

Let us now look more closely at the details of this classifier, building on our earlier introduction to the concepts.

7.2.5.1 Parametric representation of statistical data

In one dimension the statistical data for a homogeneous training field may be visualised as a single-peaked histogram having some finite spread. If this finite spread covers p data values in this one dimension then there is a need to store p values to represent the statistical histogram. If the database is extended to n dimensions, p^n storage locations are required. Putting this into perspective: previously we suggested that there were 65 536 ground cover classes that could possibly be spectrally resolved from four-band Landsat data where each ground cover class was described by four levels in each band. If the same limit of four levels is applied here then $4^4 = 256$ storage locations are required to represent the statistics for each class, when represented by an occurrence histogram. This figure may then be multiplied by the number of user specified classes.

If however the occurrence histograms may be represented by a simple mathematical function the number of parameters may be reduced. Returning to the one-dimensional unimodal histogram it is usually found (Swain and Davis 1978) that the occurrence histograms for homogeneous classes may be well represented by a normal (or Gaussian) probability density function. As such, in one dimension, the representative probability function may be described for one class by only two parameters—the mean and the standard deviation.

Constraints on representing class statistics by the (parametric) normal probability density function. Any representation of the actual class statistics will contain less detailed information than the original occurrence histogram statistics. For mathematical tractability it is convenient to simplify the original statistics by some parametric function. The question of the applicability of this function to actuality must now be considered.

So far we have confined our detailed attention to one dimension—the univariate case. We have hinted that the arguments may be extended to *n* dimensions—the multivariate case. (This extension will be made mathematically in the next section.) Further, we have indicated that the spectral signature should not be considered as a sequence of unrelated single-dimensional slices but rather as a composite variable. Similarly the extension of the probability function from the simplicity of a single dimension to multiple dimensions permits the spectral signature to be more adequately described. Accordingly discrimination between 'classes' could be expected to become more reliable.

Also, as a parametric representation is involved it necessarily renders the function less sensitive to minor violations of the homogeneity criterion. If there are enough minor violations confusion between specified classes can result as the parametric representations will tend to overlap more. An adherence to homogeneous training fields at the level of classification decided upon (Chapter 3) can remove this problem.

At this stage within the development of the parametric representation of training field statistics it is appropriate to consider the relationship between a homogeneous ground cover class and the unimodal distribution that represents it. For a series of pixels that constitute a training field for a single ground cover class, and where the class is assumed to be uniformly spread over the training field, there will be differences between the radiance values for each pixel in each channel. These differences could be produced by factors at a level of classification more refined than that regarded as being appropriate to the class. For example: soil moisture, farming practice and like effects may influence the returns for the set of pixels nominally representing a homogeneous ground class of a specific type of wheat. Similarly the passage of wind fronts over the crop can cause changes in the reflectance between pixels. These variations will then cause departures from the 'delta-function' type distribution that may at first be assumed to typify the homogeneous ground cover class. Some of these factors will already have been considered by the astute analyst in deciding the level of classification (Chapter 3) appropriate to the study. Consequently the distribution is expected to tend more towards the Normal Distribution and away from the Delta Function Distribution. Such a Normal Distribution will usually be unimodal at the level appropriate to the homogeneity level decided upon by the analyst. Perturbations to this homogeneity, such as indicated above, should be seen as minor variations upon the surface of the unimodal distribution for the class, at the level decided upon. If this is not true then obviously the division into subclasses, consistent with the appropriate level, should be contemplated.

From a purely operational standpoint: the greater the complexity of the probability density function representation, the greater is the cost of classifying the area required—in terms of computer storage and time. It has usually been found that the assumption of a normal probability density distribution provides acceptable accuracy at reasonable classification cost (Swain and Davis 1978).

Two major constraints however apply. The first is the need for the original data distributions to be unimodal. The second is the need for an adequate number of samples to be taken in order to support the statistical representation of the class.

If the unimodality of the original data distribution is violated then it is invalid to represent that 'class' by a unimodal probability function. Rather, consideration should be given to dividing the (supposedly homogeneous) 'class' into truly homogeneous, and hence statistically unimodal, sub-classes. (This may be accomplished by viewing the data histograms for each channel for each class then assessing any potential for multimodal 'classes'.)

If we consider the ratios of unimodally distributed original data bands, then these ratios may not necessarily possess unimodal characteristics (Swain and Davis 1978). However, the resultant parametric representation of the ratioed channels *is* unimodal—being defined by the applicable means and standard deviations of the derived channels. The extent by which any such resultant multimodality diminishes class separation may be assessed by allowing a Separability Index module to rank the channels according to the effectiveness of class separation. Only those channels that increase class separability should then be used in the classification process—in the interests of the cost-effectiveness of the process.

The second major constraint concerns the need for an adequate number of samples to be included in the total statistics for each class. As will be evident in the next section, at least $(n + 1)$ samples must be provided for each class, otherwise the variance/covariance matrix will be singular (i.e. its determinant will be zero and the matrix will not be able to be inverted). This would make it impossible to derive the n-dimensional probability density distribution function.

n is the number of channels in the classification. Another use of the Separability Index function is to reduce the number of channels to a 'useful' level. The number of samples possible for a class may dictate the number of channels ultimately selected. In practice it is suggested that at least $10n$, and preferably $100n$, pixels be sampled for each class (Swain and Davis 1978).

7.2.5.2 *The basis for the Maximum Likelihood classifier*

The multidimensional normal probability density function. In this section the idea of representing an occurrence level histogram for a homogeneous class, in one dimension, by a parametric normal probability density function, is extended to the similar representation in *n*-dimensional space. It is obviously impossible to insert *every* occurrence of a ground cover class over an area on the ground into the statistics for a class. (To do so would imply that the required classification had already taken place!). Accordingly 'best estimates' of the parameters that describe the class are used rather than the actual values. The more representative the sample, the better this 'best estimate' will be.

The following are the symbols used in subsequent discussions:

$p(x_{km}\|i)$	probability of a pixel at the point x_{km} in the *m*th dimension of feature space being a member of class *i* given the condition that the class *i* has been previously defined in that dimension
n	total number of channels
m,h	specific channel indices
g,i,j	specific class indices
k	specific pixel index
x	pixel
q_i	total number of pixels in class *i* sample
u_{im}	mean for class *i* for channel *m*
s_{im}	standard deviation for class *i* for channel *m*
v_{imm}	variance for class *i* for channel *m*
v_{imn}	covariance for class *i* between channels *m* and *n*
$\mathbf{X}_k$	data vector for pixel *k* in each data channel
$\mathbf{U}_i$	mean vector for class *i* over all data channels
$\mathbf{V}_i$	variance/covariance matrix for class *i* over all data channels
$\|\mathbf{V}_i\|$	the determinant of the variance/covariance matrix
$\mathbf{V}_i^{-1}$	the inverse of the variance/covariance matrix
$(\mathbf{X}_k-\mathbf{U}_i)^{\mathrm{T}}$	the transpose of the vector $(\mathbf{X}_k-\mathbf{U}_i)$ (the transpose of a column vector is a row vector).

Returning to the single-dimensional case, in channel *m*, the probability that a pixel x_k will be classified into class *i* is given by equation (7.1). This assumes a normal probability density distribution function is applicable to this case in one dimension.

$$p(x_{km}|i) = \frac{1}{(2\pi)^{1/2}s_{im}}\exp\left(-\tfrac{1}{2} \times \frac{(x_{km} - u_{im})^2}{s_{im}^2}\right) \qquad (7.1)$$

where

$$u_{im} = \frac{1}{q_i}\sum_{k=1}^{q_i} x_{km} \qquad (7.2)$$

and

$$s_{im}^2 = \frac{1}{q_i - 1}\sum_{k=1}^{q_i}(x_{km} - u_{im})^2 \qquad (7.3)$$

(from Swain and Davis 1978, p 149). (The above quantities u_{im} and s_{im} are the 'best estimates' for the true values. The relation thus relies upon a large and representative sample.)

If we now extend the concept to n dimensions—the multivariate case—we have the expression (7.4), in matrix form. Here the probability of a pixel x_k, taken over all n channels, being allocated to class i is $p(X_k|i)$.

$$p(X_k|i) = \frac{1}{(2\pi)^{n/2}|V_i|^{1/2}} \exp\left[-\tfrac{1}{2}(X_k - U_i)^{\mathrm{T}} V_i^{-1}(X_k - U_i)\right] \qquad (7.4)$$

where

$$X_k = \begin{bmatrix} x_{k1} \\ x_{k2} \\ \vdots \\ x_{km} \\ \vdots \\ x_{kn} \end{bmatrix} \qquad U_i = \begin{bmatrix} u_{i1} \\ u_{i2} \\ \vdots \\ u_{im} \\ \vdots \\ u_{in} \end{bmatrix}$$

and

$$V_i = \begin{bmatrix} v_{i11} & v_{i12} & \cdots & v_{i1n} \\ v_{i21} & v_{i22} & \cdots & v_{i2n} \\ \vdots & \vdots & v_{ihm} & \vdots \\ v_{in1} & v_{in2} & \cdots & v_{inn} \end{bmatrix} \qquad (7.5)$$

with u_{im} and v_{ihm} being given by equations (7.6) and (7.7):

$$u_{im} = \frac{1}{q_i}\sum_{k=1}^{q_i} x_{km} \qquad m = 1, 2, \ldots, n \qquad (7.6)$$

$$v_{ihm} = \frac{1}{q_i - 1}\sum_{k=1}^{q_i}(x_{kh} - u_{ih})(x_{km} - u_{im}) \qquad \begin{matrix} h = 1, 2, \ldots, n \\ m = 1, 2, \ldots, n. \end{matrix} \qquad (7.7)$$

Equation (7.4) thus allows the probability of a pixel being classified, over all n bands, into class i or j to be evaluated after reference to the set of parameters describing the probability functions for each class. This will only be possible if the inverse of the variance/covariance matrix can be produced (i.e. V_i^{-1} validly exists). (Note also that the exponent term in equation (7.4) is equal to one half of the negative of the previously presented Mahalanobis distance.)

Need for (n+1) samples to support the Maximum Likelihood classifier. We should now establish the conditions that prevent the inverse matrix V_i^{-1} from existing since they have a fundamental bearing on the operation of a Maximum Likelihood classifier. Consequently they influence the manner in which the ground truth sampling programme is conducted since the minimum number of required samples must continually be borne in mind.

The inverse of the square matrix V ($n \times n$, rows by columns) may be computed in the following way.

If V, for illustrative purposes, is a 3×3 matrix

$$V = \begin{bmatrix} v_{11} & v_{12} & v_{13} \\ v_{21} & v_{22} & v_{23} \\ v_{31} & v_{32} & v_{33} \end{bmatrix}$$

then the cofactor for v_{11} is $+(v_{22}v_{33} - v_{32}v_{23})$ and the cofactor for v_{21} is $-(v_{12}v_{33} - v_{32}v_{13})$. That is, the cofactor c_{de} for element v_{de} is given by the determinant obtained by the omission of the dth row and the eth column of V being muliplied by $(-1)^{d+e}$ (Frazer *et al* 1950). If we now prepare a matrix of the cofactors we have

$$C = \begin{bmatrix} c_{11} & c_{12} & c_{13} \\ c_{21} & c_{22} & c_{23} \\ c_{31} & c_{32} & c_{33} \end{bmatrix}.$$

The adjoint matrix V^A to V is prepared by transposing the elements in the above cofactor matrix so that $v_{de}^A = c_{ed}$, thus

$$V^A = \begin{bmatrix} v_{11}^A & v_{12}^A & v_{13}^A \\ v_{21}^A & v_{22}^A & v_{23}^A \\ v_{31}^A & v_{32}^A & v_{33}^A \end{bmatrix} = \begin{bmatrix} c_{11} & c_{21} & c_{31} \\ c_{12} & c_{22} & c_{32} \\ c_{13} & c_{23} & c_{33} \end{bmatrix}.$$

The inverse matrix V^{-1} may now be derived from the adjoint matrix by dividing the elements of the adjoint by the determinant of V, as follows:

$$v_{de}^{-1} = v_{de}^A/|V| = c_{ed}/|V|.$$

Consequently the inverse cannot usually be produced if $|V|$ is zero. (Frazer *et al* (1950) put forward a transferral technique for singular matrices—but see a later comment in this section on the number of samples.) If $|V| = 0$ the matrix V is said to be singular. Consequently the variance/covariance matrix V_i must be a non-singular square matrix if equation (7.4) is to be solved.

Several conditions can produce a zero determinant. If two, or more, columns or rows are identical then $|V| = -|V|$ and hence $|V|$ must be zero as the only definite solution (Aitken 1951).

If we turn now to equations (7.6) and (7.7) we note that the number of different samples must be equal at least to the number of channels. Otherwise a row, or column, will be all zeros—by default—and the determinant will be zero.

Let us consider the case where the number of samples q_i is equal to the number of channels, in terms of equation (7.3). We have a finite number of samples (q_i) for each class each of which has a value for the mth band, up to n. The x_{km} can take on any value out of the dataset q_i for that mth band. Consequently the first, second, third, . . ., (q_i-1) terms in the summation of equation (7.3) can also take any value derived from the permitted dataset. However the q_ith term is fixed since $\Sigma_{k=1}^{q_i}(x_{km} - u_{im})$ must equal zero. Consequently the number of 'degrees of freedom' is (q_i-1) for this finite dataset being sampled without replacement (Brandt 1970). As here q_i has been set to n, the number of samples must be increased by one to avoid $|\mathbf{V}|$ being set to zero by the above degrees-of-freedom constraint on the sampling for variance/covariance terms. Thus q_i must be equal to or greater than $(n + 1)$. (Ideally q_i should be at least $10n$ or $100n$ or more (Swain and Davis 1978).)

This then imposes a project design constraint on the *minimum number of pixels that must be included in each class*—derived from homogeneous training fields.

A valid situation can also arise leading to a zero determinant. If a class is so pure that all the pixels have identical values in one or more data channels (e.g. deep pure unperturbed water recorded by Landsat MSS 7) then at least one row/column in the variance/covariance matrix will be zero with the determinant then being zero. The software should always report this case to the user and take appropriate action, perhaps by setting a quantity to an acceptably small value (e.g. Beach 1980).

Consequently at least $(n + 1)$ *different* samples must be taken to permit $\mathbf{V}$ to be inverted and equation (7.4) to be evaluated.

7.2.5.3 Relation between probability density functions and the likelihood function

Until now we have almost exclusively considered the probability density function and not the likelihood function upon which the maximum likelihood classifier is based. The probability density and likelihood functions are not equivalent. Rather the likelihood function is derived from the probability density function. This section outlines this vital relationship between the two functions, and the differences between them.

Probability foundations. Firstly, we need to consider some basic concepts in probability. Probability exists before a measurement is taken, or a pixel classified. It is not a quantity that materialises upon completion of the measurement or the classification. This quantity, upon completion of the experiment, is known as the likelihood and is dependent upon the number of repeat experiments conducted with the same sets of probability density functions. Before discussing the step between probability and likelihood let us look at two options for probability functions that are encountered in classification.

If we consider two probability density distributions for two classes *A* and *B* as shown in figure 7.3 then the two classes are regarded as being mutually exclusive. This is depicted with reference to two-dimensional space. An equal value of the probability function is assumed distributed uniformly within the bounding contour.

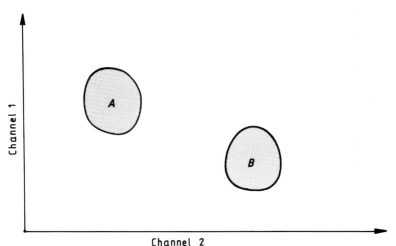

Figure 7.3 Two mutually exclusive probability density distributions for two classes *A* and *B*. A nominal 1% boundary to the distributions is assumed.

For mutually exclusive classes the probability of a pixel being classified into one of *A* or *B* is $P(A + B)$ and

$$P(A+B) = P(A)+P(B)$$

where $P(A)$ and $P(B)$ are the probabilities of the pixel being classified into either the *A* or *B* classes.

A more usual case is where the two probability density distributions overlap. That is, classes *A* and *B* are not mutually exclusive. This is depicted in figure 7.4, and follows the treatment by Brandt (1970).

If a pixel is located within the overlap region it has a probability $P(AB)$ of being a member of *both* classes *A and B*. We now ask, for a pixel within the class *A*, what is the probability of it also being a member of class *B*. This probability of the pixel being in class *B* whilst a member of class *A* is defined as the conditional probability $P(B|A)$. On reference to figure 7.4 this conditional probability can be seen as the following ratio—here diagrammatically expressed as areas:

$$P(B|A) = P(AB)/P(A).$$

If A and B are mutually exclusive classes $P(AB) = 0$ and hence the conditional probability $P(B|A) = 0$. Similarly if A is contained within B then $P(B|A) = 1$.

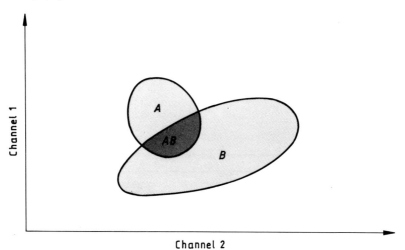

Figure 7.4 Two overlapping probability density distributions for classes A and B. A nominal 1% boundary to the distributions is assumed.

So far we have considered the probability discussion based on a single measurement—as is the case with classifying pixels individually. To indicate a conceptual difference between the probability density and likelihood functions we need to extend the theory to cover repetitive measurements.

If two measurements C and D are independent then the knowledge that C has occurred will not change the probability functions for D and vice versa. These probability functions will generally overlap to some extent as they will usually be related to repetitive independent measurements of the same quantity. Consequently, the following relation will apply:

$$P(D|C) = P(D).$$

Then from the earlier relation

$$P(D|C) = P(CD)/P(C) \qquad \text{we have} \qquad P(CD) = P(C)P(D).$$

Hence for N repetitive independent measurements of the same quantity D the resultant 'probability' Pa will be $Pa = (P(D))^N$.

Pa however is an 'after-the-experiment-is-over' type probability or an *a posteriori* probability. Following Brandt (1970) it is also termed the 'likelihood' L. The point to note is that the resultant likelihood L is derived from the completion of a series of N measurements which have all taken place, independent of one another, with the same probability of obtaining an individual result $P(D)$. The resultant likelihood of *each* of the N events yielding the *same* result is then the product of the individual probabilities of that event occurring as the result of a single independent measurement. This summarises the fundamental difference in concept between a probability density function and a likelihood function.

Let us illustrate the difference with an example as used by Brandt (1970).

'An asymmetric coin is tossed a number of times. From the result we attempt to decide whether it belongs to class A or class B. These classes have the following probability behaviours:

	Class A	Class B
Heads	$\frac{1}{3}$	$\frac{2}{3}$
Tails	$\frac{2}{3}$	$\frac{1}{3}$

On performing the experiment, five tosses yield one head and four tails. Remembering that the probability of an individual event does not change, the resultant likelihoods L are:

	Class A	Class B
Likelihood	$L_A = \frac{1}{3}(\frac{2}{3})^4$	$L_B = \frac{2}{3}(\frac{1}{3})^4$
	$= 0.06584$	$= 0.00823$

Consequently it is eight times more likely that the coin belongs to class A' (Brandt 1970).

Note again that the individual probabilities have not changed but that the resultant likelihood is a function of the number of experiments.

So far we have considered *a posteriori* probability. We now need to underline the difference between this 'after-the-event' resultant probability and *a priori* probability. A priori probability is a weighting factor applied to the single event probability to reflect other influences, e.g. spatial associations, seasonal influences etc.

For example: when classifying individual pixels it is possible for a pixel to be classified as wheat within a bare ground field. However, at that time of the year it may be more reasonable for the pixel to be classified as dry pasture. Accordingly, a set of *a priori* weightings could be applied to the individual class probabilities to yield, after the experiment (classification), a more representative *a posteriori* probability or likelihood for that pixel.

A spatial example could be: the white sand along an ocean beach may have a similar spectral signature to snowpack; yet in summer, it would be unlikely for snowpack to be by the ocean edge. In this case the analyst may enter the following *a priori* weighting factors for the region by the ocean:

$$p(\text{sand}) = 0.80$$
$$p(\text{snow}) = 0.20$$

and in the mountains
$$p(\text{sand}) = 0.20$$
$$p(\text{snow}) = 0.80$$

where $p(X_k|\text{sand}) = p(X_k|\text{snow})$ for both locations.

Two classifications would be executed with different *a priori* weightings: one for the ocean edge and one for the mountains.

Let us explicitly introduce the concept of the 'likelihood ratio', with reference to the task of classification. If a pixel *is classified into one* of a set of classes *it is not classified into any of the remaining classes*. Thus for each class a 'yes/no' type result can be computed. This is the 'likelihood ratio' for each class (Eisenhart *et al* 1947). Consequently for each pixel a set of likelihood ratios, one for each of the user specified classes, can be computed. The maximum of the set then indicates the most likely class for that pixel. Momentarily referring back to the asymmetric coin example: the two likelihood ratios were

$$L_A/L_B = 8.000 \qquad L_B/L_A = 0.125.$$

Thus class A was the more likely result.

The above discussion has only dwelt implicitly on mutually exclusive events. Overlap in the probability density functions invokes the concept of conditional probability. As we saw earlier this permits a relevant probability density function to still be derived for such non-mutually-exclusive events.

Computationally, the introduction of a logarithmic likelihood ratio can lessen the use of computer time.

Before proceeding let us recap the fundamental difference between a probability density function and a likelihood function. This is that the likelihood function is derived after a *series* of independent events from the *individual* probability density functions. The importance of this difference is evident in the derivation of the likelihood ratio.

As usually used in classification, the likelihood function is evaluated for each pixel over the set of user specified classes. The maximum value of this function, rather than the ratio, is then determined and the most likely class ascribed to the pixel. Within the derivation of the likelihood function is embedded the allowance for minimising misclassification of the pixel into other classes.

The Divergence module, see later, uses the likelihood ratio as the key relation in determining the separability between classes.

7.2.5.4 *Applying the general concept of Maximum Likelihood to classification*

We have already indicated that the decision as to the relevant class for a pixel is based on the maximum value of the set of likelihood functions for the pixel rather than the set of likelihood ratios. This uses less computer resource per pixel.

A further refinement to the general Maximum Likelihood concept is that of classifying each pixel only once. (The preceding development of the likelihood concept has concentrated on multiple trials for clarity.) In this case the likelihood function can approach the probability density function. The conceptual distinction between the two is however still maintained.

In classification it is usual to employ a 'yes/no' decision rule as to whether the pixel is correctly classified or not. This carries over in the theory to a zero–one loss function (see later). On including *this* loss function the likelihood function evaluated for a pixel for a specific class approaches the probability density function, for the same conditions, even more closely.

The above two sections are a summary of the concept of, and the constraints on, the Maximum Likelihood classifier. We must now substantiate these ideas mathematically.

7.2.5.5 *Mathematical background to the Maximum Likelihood classifier*

The multivariate probability function, $p(X_k|i)$, introduced in § 7.2.5.2 is the probability of a pixel X_k (over the n data channels constituting a measurement vector X_k) being classified into class i, given the condition that the class i has been previously defined. Obviously there is a set of probability functions $p(X_k|i), p(X_k|j), \ldots, p(X_k|r)$ covering the number of classes, r, that the analyst introduces.

To start us thinking about classification let us consider a simple case. Such a decision rule for classification based on the probability density functions could be: that X_k is allocated to class i if, and only if

$$p(X_k|i)\, p(i) \geqslant p(X_k|j)p(j) \qquad \text{for all } j = 1, 2, \ldots, r. \qquad (7.8)$$

Here $p(i)$ is the *a priori* probability associated with class i for that location and $p(i)$ is independent of class training data.

This is one of the simplest decision rules. It is based only on the probability density functions and makes no attempt to progress that next step and maximise the likelihood of correct classification. Let us consider how this extension can be made.

If we think back to the concept of the 'lumpy custard' representing a three-dimensional probability distribution and concentrate on the classification 'fate' of an element of custard located between class 'lumps' we can relate the situation to equation (7.8). These elements between 'lumps' may be confronted with equal probability function values from one or more classes. Only one class may be correct for that element although more than one may be incorrect, particularly if the exact requisite class has not been specified in the user defined set. If this correct class hasn't been included then the element will have a range of perhaps equal probabilities of being classified into different specified classes. The classifier is faced with a dilemma.

The task is now to outline a decision rule that reduces this chance of misclassification to a minimum. This decision rule is the heart of the Maximum Likelihood classifier and is known as the Bayes optimal classification strategy. It is based upon the probability density functions that have already been developed.

Let $d(i|j)$ represent the amount of loss produced if a pixel, which really belongs to class j, is mistakenly classified into class i. This set of loss functions ranges over all classes:

$$d(i|j) \qquad i = 1, 2, \ldots, r \qquad j = 1, 2, \ldots, r \qquad (7.9)$$

and is here given the values of

$$d(i|j) = \begin{cases} 0 \\ 1 \end{cases} \quad i \begin{cases} = j \\ \neq j. \end{cases} \qquad (7.10)$$

This set is then known as a zero–one *loss* function (Swain and Davis 1978). Thinking back to § 7.2.5.3 we remember that we introduced a similar *success* function by indicating that if a pixel was of one class it couldn't be of another class—a 'yes/no' decision. Equation (7.10) is a mathematical representation of the similar *loss* decision rule. As may be seen it is particularly appropriate to the task of classifying pixels into one of a set of user nominated ground cover classes.

We now need to quantify the amount of loss brought about by a pixel being incorrectly classified. The pixel X_k will usually be located amidst overlapping probability density functions. That is the classes $g = 1, 2, \ldots, r$ will not be mutually exclusive. Consequently there will be a finite probability that X_k, whilst a member of class g, will also have the characteristics of class i, see figure 7.5. In terms of deriving the amount of loss brought about by the incorrect classification of X_k into class i rather than class g we need to consider the conditional probability for X_k into class g. This is $p(g|X_k)$. We will consider each of the incorrect i classes in turn. The amount of loss for one class g, where g is used as a running index, is $d(i|g)p(g|X_k)$.

Remember that

$$d(i|g) = \begin{cases} 0 \\ 1 \end{cases} \quad i \begin{cases} = g \\ \neq g. \end{cases}$$

Thus if $M_{X_k}(i)$ is the loss resulting from pixel X_k being classified into class i

when it should be allocated to class g then

$$M_{X_k}(i) = \sum_{g=1}^{r} d(i|g)p(g|X_k).$$ (7.11)

Here $p(g|X_k)$ is the conditional probability that X_k is from class g, given that X_k exists and the probability is being evaluated of it being allocated to class g.

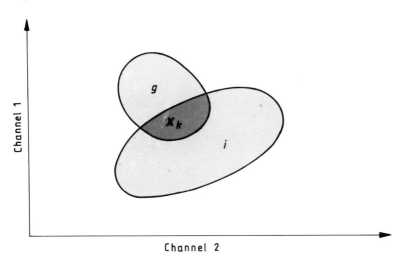

Figure 7.5 Two-dimensional representation of overlapping probability density functions and the loss dilemma that faces pixel X_k which is a member of class g whilst possessing most of the characteristics of class i.

This conditional probability is here given by

$$p(g|X_k)p(X_k) = p(X_k|g)p(g)$$ (7.12)

($p(g)$ is the *a priori* probability being retained for completeness).

$p(X_k)$ is the total probability of X_k being classified into one of the r classes and is evaluated from

$$p(X_k) = \sum_{g=1}^{r} p(X_k|g)p(g).$$ (7.13)

Note that if the class accurately described by X_k has not been completely specified, then $p(X_k) < 1.0$.

Returning to equation (7.11) and noting the loss function of equation (7.10), we see that $M_{X_k}(i)$ is a 'normalised' loss with contributions over all classes except when $i = g$ (in this maximum likelihood case).

Equation (7.12) may be substituted into equation (7.11) to give

$$M_{X_k}(i) = \sum_{g=1}^{r} d(i|g)p(X_k|g)\frac{p(g)}{p(X_k)}.$$

If we now apply the zero–one loss function (7.10) we have

$$M_{X_k}(i) = \sum_{\substack{g=1 \\ g \neq i}}^{r} p(X_k|g)\frac{p(g)}{p(X_k)}$$

$$= \frac{1}{p(X_k)}\sum_{\substack{g=1 \\ g \neq i}}^{r} p(X_k|g)p(g). \qquad (7.14)$$

Now equation (7.13) may be reduced to

$$p(X_k) = p(X_k|i)p(i) + \sum_{\substack{g=1 \\ g \neq i}}^{r} p(X_k|g)p(g).$$

The summation may thus be substituted for in equation (7.14) to give

$$M_{X_k}(i) = 1 - p(X_k|i)\frac{p(i)}{p(X_k)}. \qquad (7.15)$$

$M_{X_k}(i)$ is the loss resulting from misclassifying X_k, which was really of class g into class i. A set of values $L_i(X_k)$ can thus be set up for the pixel X_k, one for each of the specified classes. If this set of values $L_i(X_k)$ is set equal to the negative of the loss, for all classes, as follows:

$$L_i(X_k) = -M_{X_k}(i) \qquad i = 1, 2, \ldots, r \qquad (7.16)$$

then rather than minimising the loss M, an equivalent approach would be to maximise the negative of the loss functions. That is, the likelihood of correct classification is maximised. Thus, on revising equation (7.15) with equation (7.16) we have

$$L_i(X_k) = p(X_k|i)\frac{p(i)}{p(X_k)} - 1. \qquad (7.17)$$

Obviously when attempting to assess the relative likelihood levels between classes numerical constants may be omitted. Similarly, for a given X_k, $p(X_k)$ will be constant over the set of $L_i(X_k)$. Consequently equation (7.17) reduces to

$$L_i(X_k) = p(X_k|i)\, p(i) \qquad i = 1, 2, \ldots, r \qquad (7.18)$$

which forms a set of likelihood values for X_k being classified into the set of r classes specified by the user.

By reference to equation (7.4) of a preceding section the set of likelihood values becomes

$$L_i(X_k) = \frac{p(i)}{(2\pi)^{n/2}|V_i|^{1/2}} \times \exp\left[-\tfrac{1}{2}(X_k - U_i)^{\mathrm{T}}V_i^{-1}(X_k - U_i)\right]$$
$$i = 1, 2, \ldots, r. \qquad (7.19)$$

The decision rule equation (7.8) may now be rewritten as a Maximum Likelihood decision rule: where X_k is allocated to class i, if and only if

$$L_i(X_k) \geq L_g(X_k) \qquad \text{for all } g = 1, 2, \ldots, r. \qquad (7.20)$$

This Maximum Likelihood decision rule rests on:

(i) the Bayes optimal strategy of minimising the average loss due to misclassification over *all* classes. That is, this approach takes into account the probability contributions from all classes for misclassification of the pixel;

(ii) the zero–one loss function as a special case. This is the Maximum Likelihood decision rule and produces results which have the minimum probability of error over the entire classified data set. That is, this approach produces, on average, the most accurate classification (Swain and Davis 1978).

The zero–one loss function can be replaced by another function. This would change the classification type from that special subset of the Bayes optimal strategy known as the Maximum Likelihood decision rule. Such a change could be necessary if some classes had a low probability of occurrence (i.e. $p(i)$ was low). A low $p(i)$ could lead, in the zero–one loss case, to very small values for $M_{X_k}(i)$ for that class (equations (7.10) and (7.14)) with the possible resultant exclusion from the total classified dataset. A more complicated loss function can be used to overcome this case.

A priori probabilities are usually not reliably known. The usual practice is to assume that the *a priori* probabilities for all classes are equal. This results in an equal constant percentage $p(j) = P$ being allocated to each class. $p(j) = P$ depends upon the number of classes. This approach means that the discriminant function reduces to:

$$L_i(X_k) = p(X_k|i) \qquad i = 1, 2, \ldots, r \qquad (7.21)$$

and that the loss functions become equivalent to

$$d(i|j) = \begin{cases} 0 \\ 1/p(j) \end{cases} \quad i \begin{cases} = j \\ \neq j \end{cases} \qquad (7.22)$$

(from Swain and Davis 1978).

If $p(j)$ were a variable, the loss caused by an incorrect classification would be inversely proportional to the *a priori* probability of the correct class. (This could happen if one or more, but not a complete set, of *a priori* probabilities were entered into the classification process.)

In this case the Bayes optimal strategy favours the least frequently occurring classes.

However when equal $p(j)$ are entered for all classes the above implications of this strategy are not realised and the actual loss function used is

$$d(i|j) = \begin{cases} 0 \\ 1/P \end{cases} \quad i \begin{cases} = j \\ \neq j \end{cases} \qquad (7.23)$$

where P is a constant for a given classification. Hence no class favouring should occur within analyses using the default (equal $p(j)$) values for the *a priori* probabilities.

Thus when the Maximum Likelihood classifier is run, a set of likelihood values is calculated for each pixel X_k, one likelihood value for each specified class. The pixel X_k is then allocated to that class having the greatest likelihood for that pixel. This Maximum Likelihood is then recorded, within the Classification Dataset (see Chapter 10), for that pixel together with its associated class type.

Obviously when all the probability distribution functions have low values for a specific pixel—meaning that the pixel is well removed in feature space from the user nominated class locations—there are low likelihood values stored for this pixel. Whilst each pixel is classified in the Maximum Likelihood technique, it is wrong to assume that all classifications, without further treatment, are valid representations of the actual ground cover classes over all pixels. Rather, those pixels having low likelihood values should be deleted from the displayed classified results. These pixels have been allocated to the most feasible one of the specified classes—which may be well removed from actuality. A filtering process should be invoked to remove these pixels having low likelihood values from the classified dataset. This can be done by 'thresholding'. Without this thresholding the classification process is incomplete.

Once the Maximum Likelihood values have been derived for each pixel the multidimensional data channel space collapses to the single dimension of 'likelihood space'. Each pixel has now both a likelihood value and a preferred class in this likelihood space. (All this information is stored in the Classification Output Dataset within the computer.)

The imposition of a threshold for each class upon the likelihood distribution over all pixels divides the likelihood space into regions where the likelihood values are below the threshold and 'islands' where the likelihood is equal to or above the imposed threshold. These islands contain the pixels that have been allocated acceptably to one of the specified classes. This is done by the user nominating a specific threshold for each class.

If we map the threshold contour back from the class likelihood distributions to the multidimensional class probability density functions, we note that a bounding surface has been placed about the distribution for each class (as if a shell has been placed around the 'lump' in the 'custard'). Only those actual ground cover classes that have been specified by the user will have bounding surfaces surrounding them.

Those classes, which, although they exist in actuality, have not been recognised, trained on and specified by the user, will lie below the imposed threshold level. Pixels belonging to these classes will be rejected from the user supervised classification.

The reader will recall the earlier statement about a possible number of 65 536 ground cover classes for 6-bit digitised four-channel Landsat data where each class covered four CCT levels. This is a far greater number of classes than can be usually specified in practice and a lesser number than those that exist in real life. Hence we recognise the above constraint upon those classes that exist in real life but that have not been specified by the user. These pixels have therefore been validly thresholded out of the classification result.

If the breadth of CCT levels covered by each class is increased, then a lesser number of ground cover classes may be discriminated. The original number of actual ground cover classes—usually well in excess of anything the sensor/digitising system can resolve—continues to exist. However, we now have a greater number of actual ground cover classes being represented by one system recognised class. The imposition of a threshold mapped back upon these probability density functions will now usually include a greater range of subclasses within each of the user specified major class groups.

If we return to the tighter class distributions, the classification tends more to the subclass level. It starts to discriminate between the subclasses and any that are not specified are thresholded out. The user is now confronted with more subclasses which are capable of being recognised by the systems (sensor and computing) than were specified and as a result they are validly thresholded out. The inclusion, by specification, of omitted subclasses that are important to the user (via the selection of suitable training fields) is the preferred solution to this omission—if it is important to include these subclasses in the results.

Operationally the user is encouraged to note the classification results from a 0% threshold approach and compare this classification of fields of interest with the 1% results—for example. If any desired fields have been thresholded out, a training field for that subclass/class should be entered, a test classification should be initiated and viewed using both the 0% and the user selected threshold—as before. (Since the probability density distributions for homogeneous ground cover classes rapidly approach small standard deviations as the number of data channels increases, a 1% threshold is usually sufficient to exclude most pixels that do not fall into these user specified classes. Obviously every pixel in the dataset which is to be classified would have a likelihood greater than zero associated with it by the classification process. This is because the class probability functions are unable to ever actually equal zero.)

The thresholding of extraneous classified pixels is a vital end step to the classification process. A value should be able to be inserted for all classes, or selectively on a class-by-class basis. The more homogeneous the training fields are, the more separable will be the classes and the more steeply peaked will be the probability density functions. As a result, in this case, a minimal threshold will cut out most, if not all, of the extraneously classified pixels—as we have indicated above. If the original data distributions are not homogeneous, a considerable variability will be observed with the number of pixels included within a class being markedly a function of the imposed threshold. If the decrease in a classified area between adjacent thresholds is unacceptable, it could indicate that more subclasses could have been introduced within the major class than was being initially sought in classification.

A 1% threshold usually corresponds to a pixel being removed some three standard deviations from the mean, in one dimension for a homogeneous class. In Normal Distribution statistics this indicates that there is only an approximately 0.2% probability for a pixel, of the specified class, being correctly located outside the 1% boundary in *one*-dimensional feature space.

In Chapter 13 we outline the variation in the number of pixels classified as a function of the imposed threshold value. This will illustrate the above comments on the impact of thresholding on likelihood values which have been derived from homogeneous training fields.

7.3 Class Separability Indices†

7.3.1 Introduction

We are now familiar with the basic concepts of classification. The concept of an Index of Separability is more foreign to usual remote sensing analysis projects. The New Zealand group made extensive use of the Divergence separability index and it is outlined here so that the reader may evaluate the background to the actual analysis projects where it has been used (Chapters 12, 13, 14, 15). Our understanding has been based on the work by Kailath (1967), Marill and Green (1963), Swain and Davis (1978), Wacker and Landgrebe (1972) and dicussions with Ødegaard (1979).

It is inadequate to consider the separability of classes purely from the standpoint of the Euclidean distance between the means.

Each class usually has a statistical spread associated with it through the data points in each channel or single dimension. As such probability distributions can overlap, to greater or lesser extents, the class separability becomes a function of both the separation of the means and the statistical distribution of data points, within each class, for each dimension.

---

†Some of the material included in this latter part of Chapter 7 has formed the basis for an article appearing in the *International Journal of Remote Sensing*, 'A Review of Multi-Channel Indices of Class Separability' by I L Thomas, N P Ching, V M Benning and J A D'Aguanno. This material is included with their permission.

Further, the evaluation of class separability in multidimensional space from a combination of class separabilities in single-dimension space is fallacious. Rather the covariance terms, between the dimensions, need to be considered in addition to the single-dimension measures of the distributions of the class data points—the variances—and the separations between the class means.

Of the class separability indices in common use: the Euclidean distance, the Mahalanobis distance, the Divergence, the Transformed Divergence, the Jefferies–Matusita distance, and the Bhattacharyya distance, the first two fall prey to the above theoretical objections. We shall briefly review them later for completeness only but prefer now to explore the development of an index that combines the vital features of: (i) the separation of the means, (ii) the distributions of the data in single-channel space, and (iii) the interrelation of the distributions through multichannel space. From the development of this 'baseline' Divergence index, the concepts can then be easily extended to consider the other more multidimensional indices.

In the following discussions on the various Separability Indices, we assume that all are equally accurate and effective in separating classes, unless remarked upon to the contrary, for a constant number of channels and using the same number of classes and training datasets. The objectives at this stage are to:

(i) introduce the concept of a Separability Index;
(ii) indicate the background to some major indices presently in use;
(iii) outline their mathematical derivation and relation to other indices;
(iv) present some of the various strengths and areas of limitation that need to be considered.

No attempt is made here to assess their ultimate relative accuracy as this can well be done, following a technique such as outlined by Fu (1982), once the basics of their formulation—the objective here—are understood. Such an evaluation should obviously take place under the conditions indicated above, realising too that any final assessment of the accuracy of a classification rests upon the classification process itself and not directly upon the Separability Index selection of appropriate channels. In Chapter 11, we present a technique for assessing this final accuracy of a classification.

7.3.2 *What is a Separability Index?*

As indicated above, the separability of classes is not just a function of the separation between the mean locations of the class probability distributions in n-channel feature space, but also of the distribution, in that space, of that probability.

This rests upon the assumption, explained in § 7.2.5.1, that the radiance level occurrence histogram for a homogeneous class may be well represented by a Gaussian probability density function. This was concluded there to be both a usual and reasonable parametric assumption and is the approach we continue to build upon here. However, let us note that it is but a convenient approximation to reality even though it is one that usually 'works' and is commonly accepted. As we considered in § 7.2.5.3 the probability density functions may be developed into likelihood functions and these latter distributions are the basis for our discussions on separability indices.

Likelihood distributions from classes will *always* overlap and a pixel†—a point in feature space—will be allocated to that class whose likelihood at that point is numerically greatest.

If we initially restrict ourselves to one dimension and two classes, (i,j), we can visualise the overlap of two Gaussian likelihood distributions (figure 7.6). The mean positions u_i and u_j lead to the gross separability of the classes but differing spreads in the statistics of the training fields have here led to differing spreads in the likelihood distributions. Obviously a pixel having a value between the origin and X_l will be classified into class i, and with a value greater than X_l into class j. (At X_l the minimum loss condition results in the pixel being allocated to the most likely class—see the preceding section.)

If we now make u_i and u_j identical, thereby superimposing the likelihood distributions, we can visualise that around the mean region, where $u_i = u_j$, the likelihoods will favour classification into class i. Away from $u_i = u_j$, towards zero and infinity, classification into class j becomes more likely. The question that remains is whether the two classes are in fact the same. More homogeneous training data could have resolved the dilemma by indication if class j was actually constituted by training data of other classes which, when all combined within the statistical assembly process, come close to replicating the statistics of class i. One of these less dominant statistical sub-assemblies perhaps could have been class i.

Returning to figure 7.6, we note that a region of competing likelihood distributions exists around the point X_l. For well separated classes this 'competition region' will be minimal. Consequently, if we could quantify

†Strictly we should refer to 'sample' rather than 'pixel'. A sample is that independent entity which is used to test the validity or the performance of the deduced hypothesis. In this application a picture element (or pixel) is the actual piece of the earth's surface that has been sensed by the system. We could choose our sample to replicate the characteristics in feature space of the sensed pixel. However we have here, and henceforth, regarded a 'pixel' and a 'sample' as essentially interchangeable entities as far as this discussion is concerned.

the area outside this 'competition region' under the likelihood distribution curve for each class, we would have a measure of class separability. With a small region of competition this separability measure would be high, the likelihood distributions would have minimal overlap, and the classes should be easily separated in classification. The indices we discuss here are measures of this separability.

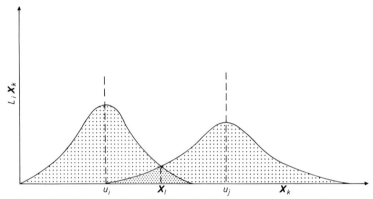

Figure 7.6 Hypothetical likelihood distributions for two classes in one-dimensional feature space.

7.3.3 The Divergence Separability Index

7.3.3.1 Derivation

Let us now evaluate this area outside the 'competition region' and hence derive an expression for the Divergence D_{ij} between a class pair (i,j) in terms of the likelihood functions such as equation (7.21), where the *a priori* probabilities for all classes are assumed to be all equal to 1.00:

$$L_i(X_k) = p(X_k|i) \qquad i = 1, 2, 3, \ldots, r. \qquad (7.21)$$

For a given point X_k the separability of the classes (i,j) depends upon the relative values of the class likelihood distributions at that point.

If we now regard this relation at the point (X_k) between the class likelihood values as a ratio, we have

$$L_{ij}(X_k) = L_i(X_k)/L_j(X_k) \qquad (7.24)$$

or, incorporating equations (7.4) and (7.21),

$$L_{ij}(X_k) = \frac{|V_j|^{1/2} \exp[-\frac{1}{2}(X_k - U_i)^T V_i^{-1}(X_k - U_i)]}{|V_i|^{1/2} \exp[-\frac{1}{2}(X_k - U_j)^T V_j^{-1}(X_k - U_j)]} \qquad (7.25)$$

where $L_{ij}(X_k)$ is the likelihood ratio between the two classes i and j at the point X_k.

If $L_i(X_k)$ is much greater than $L_j(X_k)$ at a point X_k, and X_k belongs to the ith class, then $L_{ij}(X_k)$ approaches infinity. At the 'crossover point' in the (i,j) likelihood distributions, $L_{ij}(X_k) = 1.0$, and declines thereafter into the j distribution. (If X_k belonged to the jth class the converse would obviously be true, with the crossover still occurring at 1.0.)

A glance at equation (7.25) indicates that any calculations involving likelihood ratios would be more tractable mathematically if we used the logarithmic likelihood ratio $L'_{ij}(X_k)$ (since this would employ additions/subtractions rather than multiplications/divisions) where, from equation (7.25),

$$L'_{ij}(X_k) = \ln L_i(X_k) - \ln L_j(X_k) \tag{7.26}$$

$$= \tfrac{1}{2}\ln |V_j|/|V_i| + \tfrac{1}{2}(X_k - U_j)^T V_j^{-1}(X_k - U_j)$$
$$- \tfrac{1}{2}(X_k - U_i)^T V_i^{-1}(X_k - U_i). \tag{7.27}$$

The divergence D_{ij} between the two classes i,j may now be defined as the difference between the expected value $E(L'_{ij}|i)$ of $L'_{ij}(X_k)$ when X_k is regarded as belonging to class i and the expected value $E(L'_{ij}|j)$ when X_k is regarded as belonging to class j, i.e.

$$D_{ij} = E(L'_{ij}|i) - E(L'_{ij}|j) \tag{7.28}$$

or, in terms of areas under curves,

$$D_{ij} = \int L'_{ij}(X_k)L_i(X_k)\,dX_k - \int L'_{ij}(X_k)L_j(X_k)\,dX_k \tag{7.29}$$

where the integrations are taken over the whole X_k range. The larger the difference, the greater the ability to separate the two classes.

The computations can be simplified when the classes have normally distributed density functions, by using equation (7.27). To obtain $E(L'_{ij}|i)$, we rearrange the second and third terms of equation (7.27), remembering that the expected value of X_k will be the class mean U_i. This gives

$$L'_{ij}(X_k) = \tfrac{1}{2}\ln |V_j|/|V_i| + \tfrac{1}{2}(X_k - U_i)^T(V_j^{-1} - V_i^{-1})(X_k - U_i)$$
$$+ (U_i - U_j)^T V_j^{-1}(U_i - U_j). \tag{7.30}$$

The expected value of the second term in equation (7.30) can be obtained by expansion, remembering that $(V_j^{-1} - V_i^{-1})$ is a symmetric matrix, and that, if Y is a column vector and A is a symmetric matrix, then

$$Y^T A Y = \text{tr}\, A Y Y^T \tag{7.31}$$

where tr indicates that the trace (sum of diagonal elements) of the matrix is to be taken. Thus, we find that

$$E(L'_{ij}|i) = \tfrac{1}{2}\ln (|V_j|/|V_i|) + \tfrac{1}{2} \text{tr}\, V_i(V_j^{-1} - V_i^{-1}) + \tfrac{1}{2}(U_i - U_j)^T V_j^{-1}(U_i - U_j). \tag{7.32}$$

A complementary rearrangement of equation (7.27), this time putting the expected value of X_k as the class mean U_j, enables us to find that

$$E(L'_{ij}|j) = \tfrac{1}{2}\ln|V_j|/|V_i| + \tfrac{1}{2}\operatorname{tr} V_j(V_j^{-1} - V_i^{-1}) - \tfrac{1}{2}(U_i - U_j)^{\mathrm{T}}V_i^{-1}(U_i - U_j).$$
(7.33)

Combining equations (7.32) and (7.33) we derive for the divergence

$$D_{ij} = \tfrac{1}{2}\operatorname{tr}(V_i - V_j)(V_j^{-1} - V_i^{-1}) + \tfrac{1}{2}(U_i - U_j)^{\mathrm{T}}(V_i^{-1}+V_j^{-1})(U_i - U_j).$$
(7.34)

Some texts, e.g. Swain and Davis (1978), prefer to use equation (7.31) to express the second term of equation (7.34) in trace form, namely

$$D_{ij} = \tfrac{1}{2}\operatorname{tr}(V_i - V_j)(V_j^{-1} - V_i^{-1}) + \tfrac{1}{2}\operatorname{tr}(V_i^{-1} + V_j^{-1})(U_i - U_j)(U_i - U_j)^{\mathrm{T}}.$$
(7.35)

Thus equation (7.35) consists of two terms: the first contains the differences between the respective variance/covariance matrices over n-dimensional space. The second term is the average of the Mahalanobis distances between the class means. Hence D_{ij} is never zero unless both the means and the variance/covariance matrices are identical (i.e. the classes are identical).

7.3.3.2 *Mathematical properties of Divergence*

The above derivation then leads on to five mathematical properties of Divergence (Swain and Davis 1978):

(i) $D_{ii} = 0$ The divergence of one likelihood distribution relative to itself is zero (the classes are identical).

(ii) $D_{ij} > 0$ For two different likelihood functions the divergence is always greater than 0.

(iii) $D_{ij} = D_{ji}$ Divergence is symmetrical between classes over the same n-dimensional feature space.

(iv) If the components of the composite measurement vector (the n channels of n-dimensional feature space) are statistically independent then

$$D_{ij}(x_1,x_2, \ldots ,x_n) = \sum_{k=1}^{n} D_{ij}(x_k)$$

and thus

(v) $D_{ij}(x_1, x_2, \ldots ,x_n,x_{n+1}) \geqslant D_{ij}(x_1,x_2 \ldots x_n)$.

Consequently for two different classes the addition of extra channels never decreases the class separability. (It may, however, result in a different applicable general threshold level—see the preceding section.)

7.3.3.3 *Extension from two classes to r classes*

So far we have considered Divergence only from the standpoint of separating a pair of classes. An average weighted pairwise divergence D_{ave}

may be defined for the total of r classes by equation (7.36):

$$D_{ave} = \sum_{i=1}^{r} \sum_{j=1}^{r} p(i)p(j)D_{ij}. \tag{7.36}$$

(Note that $\Sigma_{i=1}^{r}p(i) = \Sigma_{j=1}^{r}p(j) = 1$.) The interclass divergences are weighted by the *a priori* probabilities and the above relation has been based on the earlier five mathematical properties of D_{ij}.

Just as evaluating D_{ij} for a selection of different data channels allows us to determine the best set of data channels for that (i,j) class separation, then the evaluation of D_{ave} over a similar selection of data channels helps us to choose a best channel set for the differentiation of all classes, on average, one from another.

7.3.3.4 Limitations on the use of the Divergence Separability Index

Two problems, however, arise with Divergence: one with D_{ij} and the other with D_{ave}. The problem within D_{ij}, as presently formulated, occurs if the two classes are well separated and D_{ij} is evaluated from equation (7.35). As mentioned previously the second term in this equation (7.35) represents the average Mahalanobis distance between the means of the two classes. As the classes become more widely separated in feature space so does this Mahalanobis distance continue to increase. Consequently D_{ij}, evaluated from equation (7.35), continues to increase.

Once the two likelihood distributions are well separated the probability of correct (separable) classification stabilises towards 100%. However D_{ij} from equation (7.29) (or (7.35)) will continue to increase, albeit slowly. The implication is that for different sets of channels different values of D_{ij} could be obtained and their effectiveness ranked. However, once class separability is attained, this ranking should be ignored and replaced by equivalence—in terms of effectiveness in class separation.

One solution is to clip the D_{ij} value to an arbitrary upper limit. This could be the limit at which accceptable class separation is attained with the probability of correct classification approaching 100%. (Within ERMAN (IBM 1976) this arbitrary value is 999 and this clipping is the approach which is used there.)

Another solution is to transform D_{ij} into a saturating function. That is, once a certain level of separability is reached, the Transformed Divergence $^{T}D_{ij}$ remains at that (saturated) level irrespective of increases in D_{ij} brought about by the above influences.

Swain and Davis (1978) express such a Transformed Divergence in the following way:

$$^{T}D_{ij} = a[1 - \exp(-D_{ij}/b)] \tag{7.37}$$

where the constants a and b are chosen to suit the values of D_{ij} advanced by the computation process and the desired saturation value. a is selected as that saturation value and b is chosen to be at approximately the desired 'knee' position in the $^TD_{ij}/D_{ij}$ curve just before saturation (see figure 7.7, as an illustration).

Curve A in figure 7.7 presents a plot of $^TD_{ij}$ versus D_{ij} for conditions appropriate to the ERMAN system: $a = 999$, $b = 389$. (The latter constant was derived by solving equation (7.37) for $D_{ij} = {}^TD_{ij} = 900$ with $a = 999$.) As is evident, this expression for the Transformed Divergence tends to disproportionately weight the conclusions at low and high D_{ij}. This could in turn distort the determination of D_{ave}.

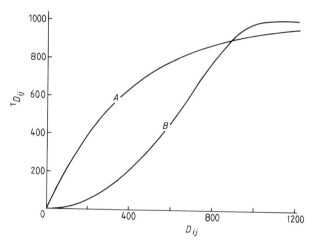

Figure 7.7 Plot of two functions suggested for the interclass (i,j) Transformed Divergence $^TD_{ij}$ as a function of the interclass Divergence $D(i,j)$, set up for the ERMAN computation conditions. Curve A, $^TD_{ij} = 999[1-\exp(-D_{ij}/389)]$; curve B, $^TD_{ij} = 999\{1-\exp[-\tan^2(D_{ij}/900)]\}$.

A more equitable expression was sought and the following is suggested here:

$$^TD_{ij} = a\{1 - \exp[-\tan^2(D_{ij}/b)]\}. \tag{7.38}$$

Again, as previously, the constants a and b are chosen to reflect the saturation value and the upper knee on the $^TD_{ij}/D_{ij}$ curve. Hence, for the ERMAN conditions a was chosen as 999 and b as 900. The resultant function is plotted as curve B in figure 7.7.

The second problem, being that with D_{ave}, occurs if two classes (i,j) are poorly separated yet are each well separated from a third class (k). D_{ij} will then be small whilst D_{ik} and D_{jk} will both be large. Due to either D_{ij} taking the clipped value or $^TD_{ij}$ taking the saturation value, D_{ave} will be large. If now (i,j) are slightly better separated and (i,k), (j,k) less well separated, we have overall better class separability but D_{ave} could be less. An incorrect deduction could therefore be made by the analyst. Two methods can be employed to overcome this problem, assuming that it is desirable to separate all classes:

(i) The selected channel sets may be ranked according to D_{ave} and the 'weak' class pairs (having 'low' D_{ij}) noted. These class pairs could then be studied individually and a channel set giving an acceptable D_{ij} noted. The final selected channel set would then be the total of those recommended by D_{ave} and by the sets of D_{ij}. This is based on the property that the addition of channels never decreases separability.

(ii) If instead of ranking the channel sets according to D_{ave} we ranked them by D_{min} we produce a best channel set that would achieve best separability for the least separable class pair. D_{min} is the value for the lowest D_{ij} for the particular set of channels selected. Again, specific sets of better channels for those (i,j) classes could be selected and added to that recommended by D_{min}.

The former approach was commonly used within the Landsat studies reported here. The latter was used in the studies on the aircraft scanner data.

7.3.3.5 Use of Divergence

Divergence, or any Separability Index, has two roles:

(i) To identify class pairs that have poor separability for a given set of data channels. This then permits a better (usually enlarged) set of data channels to be put together to improve class separability and hence raise classification accuracy.

(ii) To recommend a reduced number of channels for classification that will still produce acceptable classification results. This reduces computer time and space overheads for the project.

(The use outlined here has been almost wholly confined to the first role.)

If we now briefly digress and reflect on the number of classes ($= 65\,536$) that could possibly be discretely typified by four-channel Landsat MSS data (§ 7.2), we could suggest that some approach to acceptably reducing these possibilities would be useful. However, any such approach must still be able to recognise and include data pertinent to discriminating the selected classes required by the analyst of the dataset. The key point is that whilst an acceptable form of averaging may be invoked, if the averaging submerges the very characteristics of the original dataset that led to class discrimination, then the original dataset must be included in any final analysis.

Principal Component analysis is often regarded as an acceptable method of decreasing the dimensionality of a data set. Similarly band-ratioing can reduce total information content whilst aiding the removal of sun or sensing angle influences in the original dataset.

Suppose the original training data for each class are added to by data pertinent to the same training fields, but which have been produced from the band-ratioing and Principal Component algorithms. If the appropriate Separability Index (here the Divergence) is then invoked to suggest which are the best channels to discriminate a series of class pairs, then more evident separability can sometimes result. This is produced by using an acceptable averaging process and thus reducing what could be regarded as 'noise' in the original data—produced by the subclasses. The aim is still very definitely to improve interclass separability and this utilises other characteristics and other methods of looking at the original data to strive towards this improved separation.

Similarly the inclusion of other channels generated within a broader Geographic Information System concept could be incorporated and assessed in an identical manner with an identical objective—the better separation of classes and the resultant improvement in classification accuracy.

The usual approach to setting up and using such an index with a single acquisition data set could be advanced as follows. For the Landsat studies reported here it was customary to routinely generate a total of twenty data channels from one Landsat acquisition (being four original data channels, twelve derived ratioed channels, and four derived Principal Component channels—see Chapters 12, 13, 14). The statistics from the (homogeneous) training fields are needed to set up the mean class vectors and variance/covariance matrices. From this the mathematical base for equation (7.35) is produced.

Divergence, in the Landsat ERMAN case, was then run for all classes, selecting the best four to eight channels for overall class separations. The sets of eight, out of twenty, channels were ranked in order of effectiveness of class separation according to D_{ave}. The best set of eight channels was noted as were any 'weak' class pairs for that (average) best set of eight channels. ('Weak' class pairs were those pairs of classes whose interclass divergence D_{ij} was below a certain arbitrary figure. This was usually taken to be around 700.)

These weak class pairs were then passed into Divergence again and the best set of channels for each nominated (i,j) pair, ranked according to D_{ij}, was noted. These were then added to the previoius best overall set as a selected set for classification. (This presupposed that the analyst team was content that they wanted the classes to be separated and that the training data were homogeneous and representative.) Needless to say the Divergence studies necessarily preceded the classifications reported here.

The second role for Divergence is to provide an assessment of class separability as a function of the number of channels employed. This can reduce the use of computer resources and an example of such a study follows.

7.3.3.6 Class separability, determined by Divergence, as a function of the number of channels used

As outlined earlier, the Divergence function provides a convenient indicator of the possibility of separating classes once a classifier has been invoked. Again, class separability cannot be decreased by the addition of channels. However, as channels are added the use of computer resources is also increased. It was therefore decided to investigate the change in apparent class separability, as a function of the number of channels, to see if it was generally possible to reach a plateau region in separability where the addition of extra channels provided little return for the expenditure of resources.

The interclass pair Divergence D_{ij} was calculated for each set of class pairs for each possible set of 1, 2, 3, . . . ,6, 7, 8 channels. For each possible set of channels the average and minimum Divergence values were evaluated. This allowed the effectiveness of that channel set for class separation to be gauged and all channel sets to be ranked in effectiveness. For the reasons advanced in § 7.3.3.4, the minimum Divergence D_{min} was used for this study.

A set of class statistics to operate on the mean vectors and variance/covariance matrices was selected. These were for the 59 agricultural classes sought in the Darfield aircraft scanner study, see Chapter 15. The best 1, 2, 3, . . . ,6, 7, 8 channels were selected from the available eleven-channel database and the results are presented in table 7.1. The channel sets were ranked in effectiveness according to D_{min}—to obviate the previously mentioned truncation objection (§ 7.3.3.4). Also in table 7.1 the change in D_{min} is presented as a function of the increasing number of channels, and the increase in projected computer classification time is compared with a single-channel classification. These calculations and the resultant increases in computation time will be outlined in the next section (§ 7.3.3.7). These results are also portrayed in figure 7.8.

As seen in § 7.3.3.7, the number of calculations becomes rather large in Divergence as the requested number of channels for selection increases. ERMAN is limited to a maximum of eight and this imposes a limit on the results presented here.

For *this test example* it is noted that D_{min} reveals little increase in class separability past five channels yet computer use continues to climb unabatedly. For other sets of data (classes, acquisition systems, etc) the details could be expected to differ but the tendency towards reaching a plateau would be preserved for some specific number of channels.

Table 7.1 Class separability portrayed as a function of the number of channels selected. The channel set with the highest separability index for each selection is presented together with the separability index, D_{min} (see text). The difference in D_{min} as extra channels are added is presented (ΔD_{min}) as is the increase in computer time for a classification referred to a single-band classification as unity ($T_{F/1}$), from equation (7.42). The database used was the eleven-channel aircraft scanner data being studied with the 59 classes used in the Darfield agricultural study (Chapter 15).

No of channels	Selected set of channels	D_{min}	ΔD_{min}	$T_{F/1}$
1	9	2	—	(1)
2	8,10	118	116	3
3	3,9,10	422	304	6
4	2,4,9,10	609	187	10
5	2,4,6,9,10	738	129	15
6	2,4,6,8,9,10	763	25	21
7	1,2,3,4,7,8,10	788	25	28
8	1,2,3,4,6,7,8,10	804	16	36

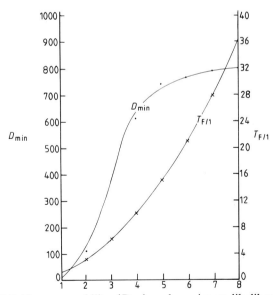

Figure 7.8 Class separability (D_{min}) and maximum likelihood classification computer time, referred to a single band classification ($T_{F/1}$) as a function of added channels. Statistics are from 59 classes sought in the Darfield agricultural programme using eleven-channel aircraft scanner data and processed via the ERMAN Divergence module.

The implications for more effective use of computer resources are clear—and this is one of the designated roles of the Divergence function, as stated earlier.

7.3.3.7 The number of calculations in Divergence and Classification

Both Divergence and Classification involve many calculations. Once a 'bench mark' run has been done for either operation, the impact of other work on available resources may be gauged. To assist in this assessment the present section gives a summary of the major numbers of calculations which may be inserted in the scaling ratios to gauge the projected use of computer resources.

Divergence. Divergence rests on the selection of a channel set from within a greater channel set (i.e. 4 channels may be selected from a set of 20). This may be accomplished in a number of ways. Once a channel set has been selected, then, from equation (7.35) the interclass divergences are computed.

The first step is to calculate the number of possible choices of the required set of channels.

Some basic terms are as follows (from Eshbach 1975).

A *permutation* of n channels is any arrangement of any number of them in a definite order. The possible number of different arrangements of n channels taken m at a time is ^nP_m where $^nP_m = n!/(n - m)!$ (where ! denotes a factorial).

A *combination* of n channels is any set of any number of them without reference to their order within the set. The possible number of different sets of n channels taken m at a time is ^nC_m where

$$^nC_m = {}^nP_m/m! = n!/(n - m)!m!. \qquad (7.39)$$

In our case we wish to select m channels from a set of n channels without reference to any ordering within the set—the ^nC_m case.

If there are r classes then the number of class pairs N_c is

$$N_c = \sum_{i=1}^{r} (i - 1). \qquad (7.40)$$

Consequently, the total number of selected operations for r classes being processed to find the best (statistically independent) set of m channels drawn from a total set of n channels is T_D, where

$$T_D = {}^nC_m \times \sum_{i=1}^{r} (i - 1)$$

$$T_D = \frac{n!}{(n - m)!m!} \times \sum_{i=1}^{r} (i - 1). \qquad (7.41)$$

(The Divergence calculations were thus often operationally referred to as ^nC_m (or $^{20}C_8$, etc) calculations.) A variety of T_D values is given in table 7.2.

Table 7.2(a) Sample *component* factors for Divergence calculations selecting m channels from a set of n channels for r classes.

Number of channels in set, n	Number of selected channels, m	Combinations of selected channels, ^nC_m	Number of classes, r	Number of class pairs, N_c
11	3	165	2	1
11	4	330	5	10
11	8	165	10	45
15	2	105	15	105
15	4	1 365	20	190
15	6	5 005	25	300
15	8	6 435	30	435
20	4	4 845	35	595
20	6	38 760	40	780
20	8	125 970	45	990
30	4	27 405	50	1 225
30	6	593 775	55	1 485
30	8	5 852 925	60	1 770

Table 7.2(b) *Total* number of selected operations for various selections of classes and required channels from the Divergence operations, computed using equation (7.41).

Number of channels in set, n	Number of selected channels, m	Number of classes, r	Total number of selected operations, T_D
11	3	55	245 025
11	4	60	584 100
11	8	60	292 050
15	4	25	409 500
15	8	25	1 930 500
20	4	15	508 725
20	4	30	2 107 575
20	8	15	13 226 850
20	8	30	54 796 950
30	4	60	48 506 850
30	8	60	10 359 677 250

Classification. The classification time is approximately proportional to $n(n + 1)$ where n is the number of channels used (Swain and Davis 1978). Consequently, once a 'bench mark' classification has been established, a proportionality scaling can take place dependent upon the change in the number of pixels to be classified and the number of channels to be used. Let T_F be the CPU time expected for the first classification. Then

$$T_F \propto \frac{P_F}{P_1} \times \frac{n_F(n_F + 1)}{n_1(n_1 + 1)} \times T_1 \qquad (7.42)$$

where P_F is the final classification number of pixels; P_1 is the 'bench mark' classification of the number of pixels; n_F is the final classification number of channels; n_1 is the 'bench mark' classification number of channels; T_1 is the CPU time of initial classification.

Performance statistics over a number of tasks on the IBM ERMAN system in Sydney, Australia, have been presented by Beach and Dawbin (1981).

7.3.4 Review of some Separability Indices

Some of the major Indices of Separability used in remote sensing analysis will now be reviewed. Consistent symbolism will be used and their properties briefly indicated. We here represent this set of Separability Indices by S.

As we indicated previously the first two are reviewed more for completeness than for their applicability to multidimensional separability investigations. However it is acknowledged that both the Euclidean and Mahalanobis Distance measures have relevance to initial evaluations since they are less demanding of computer resources, etc. If the single-channel data alone were to be studied then the Mahalanobis Distance would be considered adequate. For the final analysis, if multidimensional data are used then one of the subsequent indices should really be used.

7.3.4.1 The Euclidean Distance

In matrix notation—with the matrices as defined previously (§ 7.2.5.2)— the square of the Euclidean Distance Separability Index $_ES_{ij}^2$ between two classes (i,j) is given by the Pythagorean distance between the class means:

$$_ES_{ij}^2 = (U_i - U_j)^T(U_i - U_j). \qquad (7.43)$$

As we have found, the probability distributions for each class will be located around the mean position in multichannel space and further these distributions will usually have different 'shapes', or topology, between classes. Again, as we have previously discussed, the Euclidean Distance does not take any account of overlapping distributions. For these reasons we progressed to those functions which included the parametric index of the topology of the distribution—the variance/covariance matrices.

(Throughout we assume that the actual class occurrence distributions may be represented by a set of parameters rather than the individual points.)

7.3.4.2 The Mahalanobis Distance

The Mahalanobis Distance Separability Index $_MS_{ij}$ between two classes (i,j) is the square of the distance between the two classes expressed in terms of the variances, as indicated in § 7.3.3:

$$_MS_{ij} = (U_i - U_j)^T(V_i^{-1} + V_j^{-1})(U_i - U_j). \qquad (7.44)$$

However, this is still essentially only a transformed measure of the separability between the class means. An allowance for overlapping topology of the class probability distributions is required and this, in parametric notation, is expressed through the variance/covariance matrices.

A further practical problem in the use of the Mahalanobis Distance as a measure of interclass separability is the fact that it is unbounded. $_MS_{ij}$ will continue to increase even after 100% class separability is reached so long as the distance between the class means is increasing. The interclass Divergence has exactly this problem, for the same reason, as outlined in § 7.3.3.4.

7.3.4.3 The Divergence

Just as the Mahalanobis Distance measure was developed from the expression for the Euclidean Distance, so is the Divergence expression developed from that for the Mahalanobis Distance.

Referring to equation (7.25) we note that the Mahalanobis Distance expression appears in the exponent terms for the Likelihood Ratio

$$L_{ij}(X_k) = \frac{|V_j|^{1/2} \exp\left[-\frac{1}{2}(X_k - U_i)^T V_i^{-1}(X_k - U_i)\right]}{|V_i|^{1/2} \exp\left[-\frac{1}{2}(X_k - U_j)^T V_j^{-1}(X_k - U_j)\right].} \qquad (7.25)$$

From this was developed the expression for the interclass Divergence D_{ij} (equations (7.34) and (7.35)):

$$D_{ij} = \frac{1}{2}\operatorname{tr}(V_i - V_j)(V_j^{-1} - V_i^{-1}) + \frac{1}{2}(U_i - U_j)^T(V_i^{-1} + V_j^{-1})(U_i - U_j). \qquad (7.34)$$

As indicated in § 7.3.3, the first term in equation (7.34) contains the differences between the variance/covariance matrices over n-dimensional space and the second term is the average of the Mahalanobis Distance between the class means.

We have indicated in § 7.3.3.4 that the major practical problem with D_{ij} is that it continues to increase even after full class separability is attained. Various solutions, both theoretical and operational, were suggested in § 7.3.3.4 to overcome this disadvantage.

7.3.4.4 The Transformed Divergence

As discussed in § 7.3.3.4, a solution to the 'runaway' nature of the Divergence function is to introduce a saturating form of the function. Two forms of this Transformed Divergence $^{T}D_{ij}$ were suggested, with the first (equation (7.37)) being drawn from Swain and Davis (1978):

$$^{T}D_{ij} = a[1 - \exp(-D_{ij}/b)] \tag{7.37}$$

$$^{T}D_{ij} = a\{1 - \exp[-\tan^2(D_{ij}/b)]\}. \tag{7.38}$$

In both cases, the constants (a,b) are selected by the user to suit the computation characteristics of the system evaluating D_{ij}. As is seen in curve A, figure 7.7, the form of equation (7.37) tended to exaggerate the results at low D_{ij} and underemphasise the results at high D_{ij}. For this reason the form of equation (7.38) was advanced and this may be compared in curve B, figure 7.7.

The Transformed Divergence then is believed to be superior to the Divergence as a measure of separability. Both include, as is evident from equations (7.34), (7.35), the separability of both the means and the variance/covariance parameters.

7.3.4.5 The Jeffries–Matusita Distance

The Jeffries–Matusita (J–M) Distance Separability Index is very similar to the Transformed Divergence—particularly to the form of $^{T}D_{ij}$ expressed in equation (7.37).

The J–M Distance has been defined (as cited by Swain and Davis 1978) as

$$J_{ij} = \left(\int_{X} \{[p(X|i)]^{1/2} - [p(X|j)]^{1/2}\}^2 \, dX \right)^{1/2}. \tag{7.45}$$

This reduces, for the normally distributed probability functions we are discussing here, through equation (7.46) to equation (7.47) as has also been outlined by Kailath (1967) and Swain and Davis (1978) to

$$J_{ij} = 2\left(1 - \int_{X} [p(X|i)p(X|j)]^{1/2} \, dX \right)^{1/2} \tag{7.46}$$

and hence

$$J_{ij} = \{2[1 - \exp(-a_{ij})]\}^{1/2} \tag{7.47}$$

where

$$a_{ij} = \tfrac{1}{8}(U_i - U_j)^{\mathrm{T}}[(V_i + V_j)/2]^{-1}(U_i - U_j) + \tfrac{1}{2}\ln\left(\frac{|\tfrac{1}{2}(V_i + V_j)|}{(|V_i| \times |V_j|)^{1/2}}\right).$$

As may be seen when comparing equations (7.47) and (7.37), J_{ij} is a saturating function and is thus more appropriate as a measure of interclass separability than the Divergence. J_{ij} does however also tend to overemphasise the results for small interclass separations and underemphasise the results for the greater separations (compare figure 3.18 of Swain and Davis 1978 and figure 7.7 here).

7.3.4.6 The Bhattacharyya Distance

Kailath (1967) defined the Bhattacharyya coefficient b_{ij} as

$$b_{ij} = \int [p(X|i)p(X|j)]^{1/2} \, dX. \tag{7.48}$$

From the coefficient b_{ij} the Bhattacharyya Distance was defined as

$$B_{ij} = -\ln b_{ij}. \tag{7.49}$$

On comparing equation (7.48) with equations (7.46) and (7.47) we note that

$$b_{ij} = \exp(-a_{ij}). \tag{7.50}$$

Thus, from equations (7.49) and (7.50)

$$B_{ij} = a_{ij}.$$

Consequently for this definition of the Bhattacharyya Distance (see Kailath 1967 for references to other definitions) we have that

$$B_{ij} = \tfrac{1}{8}(U_i - U_j)^{\mathrm{T}}[(V_i + V_j)/2]^{-1}(U_i - U_j) + \tfrac{1}{2}\ln\left(\frac{|\tfrac{1}{2}(V_i + V_j)|}{(|V_i| \times |V_j|)^{1/2}}\right). \tag{7.51}$$

Fu (1982) breaks B_{ij} into two parts: the first term in equation (7.51) being due to the difference of the class means (compare this with the Mahalanobis Distance again); and the second term being due to the difference between the variance/covariance matrices.

Now as b_{ij}, from equation (7.50), can range between $+1$ and 0, B_{ij}, from equation (7.49) can range from 0 to $+\infty$. Again the separation of the class means is the major cause of this continued increase in B_{ij} after acceptable class separability is reached.

From Kailath (1967) we can see that the Bhattacharyya Distance is more appropriate to interclass separability problems than the Divergence when the class probability distributions are broad. However, when the classes are well defined both the Bhattacharyya Distance and the Divergence approaches yield similar results. (We are here assuming that the class distributions may be represented by Normal statistics with unequal means and covariances.)

The introduction of a similar saturating function to equation (7.38) for B_{ij}, for example equation (7.52), could well enable the usefulness of the B_{ij} to be extended. If $^{\mathrm{T}}B_{ij}$ is the Transformed Bhattacharyya Distance then $^{\mathrm{T}}B_{ij}$ is advanced here in the form

$$^{\mathrm{T}}B_{ij} = c\{1 - \exp[-\tan^2(B_{ij}/d)]\} \tag{7.52}$$

where c and d again are chosen to suit the computational properties of the system evaluating B_{ij}.

7.3.5 *The choice of a Separability Index*

From the preceding review (§ 7.3.4) it is evident that each of the reviewed indices has a number of advantages/disadvantages: in terms of the computer resources needed, applicability to one or more dimensions, ability to reliably assess separability, etc. Running through all of them is the desire for well-defined class distributions, such as are derived from 'homogeneous' training fields. Also the continuing increase in the index once class separation is achieved can be a problem at high separabilities—particularly if the overall best channel set is required via D_{ave} or J_{ave} or B_{ave}. Against the introduction of the relevant saturating functions $^TD_{ij}$ or $^TB_{ij}$ is the fact that any such transformation will degrade the separability over some part of the range. We are also aware, despite the above remark about 'homogeneous' training fields, that inevitably in the real problem world of the user some overlap of classes does occur. Also the constraints of the user's existent program suite, and to some extent hardware, will dictate an appropriate choice ('what can be easily incorporated/modified from what we have').

Acknowledging the above factors, and against our own experience, we would approach the choice thus:

(i) If the classes were tending towards true homogeneity and a limited channel set was in use so that the Separability Index was below the 'plateau'—see figure 7.8—our choice would be the Divergence.

(ii) If the same situation as (i) applied but the whole separability range was being used—usually due to a more extended channel set—our choice would be the Transformed Divergence.

(iii) If the classes were less truly homogeneous (i.e. perhaps the separation of classes at higher levels—see Chapter 3—was being striven for) and a limited channel set was in use—our choice would be the Bhattacharyya Distance.

(iv) If the same situation as in (iii) applied, but the full channel set was being used—our choice would be the Transformed Bhattacharyya Distance.

These would be our choices, from the Separability Indices reviewed, for a first 'cut' at the task.

However, any Separability Index is but a 'means-to-an-end'. The final choice is the selection of channels that are used in the classification process and the final evaluation is how good, or bad, is that classification product. (This latter assessment will be addressed in Chapter 11.) An approach to evaluating the effectiveness of Separability Indices on the basis of determining the upper limits to the error in separating the classes via various indices is advanced by Fu (1982).

7.4 References

Aitken A C 1951 *Determinants and Matrices* (Edinburgh: Oliver and Boyd)

Anderson J R, Hardy E E and Roach J T 1972 *A Land Use Classification System for Use with Remote Sensor Data* US Geological Survey Circular 671 (Washington, DC: USGS)

Beach D W 1980 Informal communication (IBM Sydney)

Beach D W and Dawbin K 1981 An application of large scale computing facilities to the processing of Landsat digital data in Australia *Proc. 1981 Machine Processing of Remotely Sensed Data Symp. (Purdue University, IN, USA)*

Brandt S 1970 *Statistical and Computational Methods in Data Analysis* (Amsterdam: North-Holland)

Eisenhart C, Hastay M W and Wallis W A 1947 *Selected Techniques of Statistical Analysis* (New York: McGraw-Hill)

Eshbach O W 1975 *Handbook of Engineering Fundamentals* (New York: Wiley)

Frazer R A, Duncan W J and Collar A R 1950 *Elementary Matrices and Some Applications to Dynamics and Differential Equations* (Cambridge: Cambridge University Press)

Fu K S 1982 (ed.) Application of pattern recognition to remote sensing *Applications of Pattern Recognition* (Boca Raton, FL: CRC Press)

IBM 1976 *Earth Resources–Management II (ERMAN II) User's Guide* Program No 5790-ARB, Doc. No SB11-5008-0 (Brussels: IBM)

Kailath T 1967 The divergence and Bhattacharyya distance measures in signal selection *IEEE Trans. Commun. Technol.* **COM-15** 52

Marill T and Green D M 1963 On the effectiveness of receptors in recognition systems *IEEE Trans. Inf. Theor.* **IT-9** 11

Ødegaard H 1979 Informal Communication

Swain P H and Davis S M 1978 *Remote Sensing: The Quantitative Approach* (New York: McGraw-Hill)

Wacker A G and Landgrebe D A 1972 Minimum distance classification in remote sensing *Proc. 1st Canadian Symp. on Remote Sensing* vol 2 p 577

8 Image Registration

8.1 Introduction

Image Registration involves mapping each point in an input image, which may be a Landsat sub-scene, to a corresponding point in an output image. Such an output image would conform to a recognised map projection or specified coordinate system. The mapping function may be either parametric or interpolative (linear or non-linear) (see § 8.3). In general, the size and number of pixels in the output image will not be the same as those in the input image. The process can involve image rotation, translation, and scaling or stretching.

The term 'Registration' is accepted to mean the alignment process by which two images of the same region are positioned to be coincident with respect to each other (Bernstein 1978, Hord 1982, Moik 1980, Swain and Davis 1978). The 'images' may be two remotely sensed images, such as Landsat or an aircraft scanner image, or a remotely sensed image and a map.

'Rectification' of an image is usually accepted to mean that process by which the geometry of an image is made planimetric by identifying and removing the component errors such as tip and tilt of the platform, scanner geometry etc (Bernstein 1978, Hord 1982).

We here concentrate on the concept of Registration, as the Rectification process is a necessary precursor to Registration. Every scanner/photographic/sensor system will have its own characteristics which must be compensated for in producing a rectified image. Once the planimetric image or dataset has been prepared then image to image, or image to map, registration may commence.

A more than cursory account of Registration is beyond the scope of this work. We are concentrating on the classification of remotely sensed data. Several excellent dissertations, e.g. those cited above, have been published and we assume here that 'registration' is part of the overall classification process and follows these techniques.

8.2 Registration in the Classification Process

For a classification product to be of general use it must be registered to some familiar map base. A registration *can* take place before or after classification. However depending upon the type of registration process employed the spectral data can become corrupted.

There are three main resampling techniques used in registration: nearest neighbour, bilinear interpolation and cubic convolution. (These have been ranked in order of the increasing use of computer resources and complexity.)

If a nearest neighbour resampling technique is used then the spectral characteristics of the pixel input to the process are preserved. If one of the other two is used the spectral characteristics are modified by interpolation between adjacent samples. Consequently if other than a nearest neighbour approach is used to register an image before classification, the spectral purity will be degraded for each originally sensed pixel and a less effective classification will result. If the registration is done after the classification then the allocated classes (allocated to the pixels) would be corrupted by the interpolation process. The nearest neighbour technique may be applied before, or after, a classification since sample to sample interpolation is not involved. It is however usually preferable to perform the registration following classification for single image studies. This is because the number of pixels that form the registered dataset is usually greater because of resampling, compared with the raw dataset. Obviously for multiple image classifications (say, in time sequential classifications) registration must precede classification.

8.3 An Overview of the Registration Process

In the registration process two operations take place—*geometric corrections* and *resampling*.

Geometric correction can be applied using either a parametric or an interpolative method. The former is used where known systematic errors, such as physical (earth curvature, earth rotation, relief, refraction), system (panoramic distortion, non-linear mirror sweep, scan frequency) and scanner dynamics (scanner motion, tilts of platform) can be corrected using explicit transformations. The interpolative method is where the unregistered image is warped/stretched to fit an output image using ground control points to obtain a best mean fit (Leberl 1982).

It is mainly the interpolative method which is in common use, since the user does not need to have or to calculate the sensor positional data etc, and one software package can generally be used for all types of sensor imagery. Resampling assigns multispectral values to each new pixel location. This can be achieved either by choosing the nearest neighbour or by interpolating (digital convolution) between adjoining pixels (figure 8.1). The advantages/disadvantages of sampling techniques in the classification process are outlined above in § 8.2.

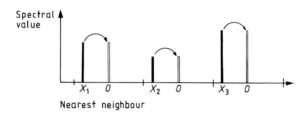

Nearest neighbour

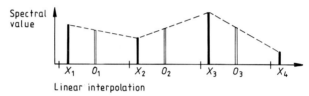

Linear interpolation

Figure 8.1 Nearest neighbour and linear interpolation resampling of raster scan data X, such as Landsat, into a new raster O.

The remainder of this section concentrates on an interpolative geometric correction with nearest neighbour resampling. It is a simple example of the registration process and is the approach used in supporting the studies reported in Chapters 12, 13, 14 and 15 and commented on by Ellis *et al* (1982).

The first procedure in the registration process is to create the reference/output grid. The latitudes and longitudes (say) of the boundaries of the image to be rectified are first determined, and these form the edges of the reference image. A reference/output pixel size for example, of 2, 4, 40, 80 etc metres ground resolution can then be chosen by the user. It is an advantage to use smaller pixels, if possible, for the registered image than occur in the input image. This is because resampling at the same pixel size can decrease the spectral resolution due to quantisation effects and can also tend to degrade the contrast ratio of ground features—if other than the nearest neighbour approach is used. As a guide, a sampling size of $1/\sqrt{2}$ of the input pixel size is considered appropriate (Mulder 1982). Hence a Landsat 79×56 m^2 pixel should be sampled at $56 \times 1/\sqrt{2} \simeq 40$ m. (An increased resampling rate may not be possible if there is a large scene to be registered, because of the large increase in data volume associated with even a modest reduction of pixel size.)

The nearest neighbour resampling process operates by taking a pixel in the reference output image and finding the pixel in the input image which most nearly corresponds to it, in terms of its location in real (latitude/ longitude) space. The reference output pixel then takes the radiance value of the pixel in the input image. The process is repeated for all pixels in the reference/output image.

The relation between pixel position in the reference/output image and the corresponding pixel in the input image is determined by a mapping function. The operator can usually choose the order of the polynomial to be used, from a first-order (linear) function up to some maximum order. The mapping function can usually either be explicitly entered by the operator, or can be determined from an input set of Ground Control Points (GCP) whose positions have been carefully identified on the input image, and whose coordinates have been accurately measured from the map. These GCP are used to find the coefficients of the mapping polynomial by a bivariate regression algorithm which produces a 'least squares' fit, or by some similar process (see Hord 1982, Moik 1980). Once the registration is completed, a further set of assessment points can be used to check the accuracy of the fit. Since the reference/output image uses latitudes and longitudes as its axes, it is a simple matter to find the linear functions which relate the predicted latitude/longitude values to the line/pixel values in the reference/output image.

A first-order polynomial is usually treated differently in the registration process to the higher-order mapping polynomials. The linear relationship between input and output pixel positions enables a relatively simple and fast algorithm to be used. Where input/output scene rotations of ten degrees or more are required the users are recommended to use the first-order polynomial function initially to achieve a partial image registration, and then to re-register this image with a higher-order polynomial function if the positional errors are still too high.

If a higher-order polynomial mapping function is used a matrix of 'anchor points' in the reference/output image is generated. The polynomial is then used to find the nearest corresponding pixels in the input image. The correspondence of pixels in the region between the anchor points is found by interpolation. This process is faster than one which uses the polynomial to compute the corresponding input pixel for every pixel in the output image.

The use of higher-order mapping polynomials should be treated with caution, since, even with a good fit of the GCP themselves, the function can become unstable between these points. The problem becomes greater, as you would expect, as the order of the function is increased (Carle and Frederiksen 1981). Even with a bivariate second-order polynomial, the number of coefficients to be determined is twelve, and an adequate number and distribution of GCP is required. Orti (1981) discusses the optimum positioning and number of these GCP in an image.

8.4 References

Bernstein R 1978 (ed.) *Digital Image Processing for Remote Sensing* (New York: IEEE Press)

Carle C and Frederiksen P 1981 *Geodetic Correction of Landsat Images—a Case Study* Dept of Surveying and Photogrammetry (Lyngby: The Technical University of Denmark)

Ellis P J, Ching N P and Benning V M 1982 Registration of Landsat imagery *Computer Classification of Landsat and Aircraft Scanner Images—The Collected Papers of ERMAN Project* ed. I L Thomas, Physics and Engineering Laboratory, Report No 776 (Wellington, NZ: DSIR)

Hord R M 1982 *Digital Image Processing of Remotely Sensed Data* (New York: Academic)

Leberl F 1982 Geometric transformation in image processing *Course in Principles of Digital Image Processing and Pattern Recognition, ITC Enschede, Holland*

Moik J G 1980 *Digital Processing of Remotely Sensed Images* publ. No SP-431 (Washington, DC: NASA)

Mulder N J 1982 Discussion *Course in Principles of Digital Image Processing and Pattern Recognition, ITC Enschede, Holland*

Orti F 1981 Optimal distribution of control points to minimise the Landsat image mean square registration errors *Photogrammetric Eng. and Remote Sensing* **47** No 1

Swain P H and Davis S M 1978 *Remote Sensing: The Quantitative Approach* (New York: McGraw-Hill)

9 An Analysis Pathway to Classification Products

9.1 Introduction

Classification is regarded as being a combination of factors: data acquisition, human decisions and software operation. The end product is a ground cover classification output which can be in one of two forms: from a lineprinter in symbol form, or on magnetic tape as decimal numbers representing the colours that are subsequently produced in a photographic transparency.

A major component in the classification process is a user-friendly system. Here a discipline-oriented user (for example, a cartographer, forester or agriculturalist) can interact directly with the computer system. This allows the user to shape his/her own analysis bringing into play other factors—e.g. planting history, disease stress, climatic/soil conditions—without the need to work through a filtering and reinterpreting computer specialist.

This user-friendliness is usually accomplished through explanatory menus written in easily understood language that may only require triggered cursor input to initiate processing actions. These menus would be supported by a flexible, but non-infinite, range of processing and display options along the analysis pathway. The further back-up support of a system consultant (spacecraft/aircraft and computer specialist familiar with the software) allows a very penetrating approach to be adopted by the user.

This chapter outlines the ERMAN analysis system and the processing pathway that was used to produce the results presented in Chapters 12–15 purely from the standpoint of one user-friendly system that could support user needs.

9.2 ERMAN Introduced

ERMAN is one of many analysis systems that can support user needs. Further, it is one of a smaller group of packages that may be described, in our opinion, as user-friendly—where by the 'user' we mean that person who doesn't want to become a computer operator, but rather use the computer as a tool to support his or her project analyses. (ERMAN has since been superseded by IBM with the Hacienda package which is now known as the IBM 7350 Image Processing System. This system has grown and broadened from the ERMAN concepts.)

9.2.1 Earth Resources Management II or ERMAN *II*

IBM (1977) introduce ERMAN II in the following terms:

'ER-MAN II consists of a large set of software programs which execute on general purpose IBM digital computers interacting with an analyst at a terminal. The purpose of the original system was to provide research analysts at the NASA Johnson Space Center with a tool to:

(i) analyse remotely sensed data;
(ii) develop and perfect remote sensing data processing techniques;
(iii) determine the feasibility of applying the data to significant earth resources problems.

'The predecessor system has been in use since September 1972. It executed on computers in the NASA Mission Control Center used in the Gemini and Apollo Manned Space Flight Program under a special operating system connected to specially designed, one-of-a-kind, terminals. In 1975, IBM undertook the task of modifying the system to execute under the control of standard operating systems and utilizing comparatively inexpensive terminals available 'off the shelf'.

'Although remote sensing data analysis can be a fairly complex task, ER-MAN II was designed to provide a very simple man/machine interface. Analysts bring the data to be processed to the computer on digital tapes and sign on at the terminal. ER-MAN prompts the analysts for inputs by presenting 'menus' on the terminal, e.g. figure 9.1 being the master control menu for the pattern recognition section of ER-MAN. The analyst controls the system and enters data by selecting multiple choice options with a cursor, or by typing in parameters. Results are displayed on a black and white alphanumeric TV screen or an image screen which can display grey shade or color images. ER-MAN checks all inputs for possible errors and displays error messages with corrective actions when one is detected.

'The system has a large complement of powerful application programs capable of performing a wide range of data analysis and manipulation functions and producing output products' (see also figure 9.2).

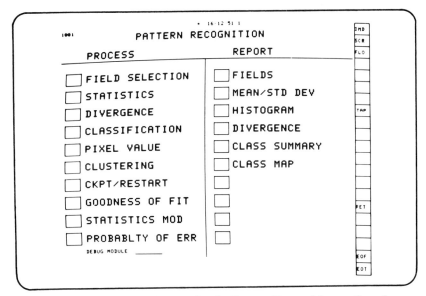

Figure 9.1 The master control menu for the Pattern Recognition section of ERMAN (IBM 1976).

9.2.1.1 ERMAN *as used by the New Zealand group*

ERMAN II was installed by IBM Australia Ltd in Sydney, and the New Zealand group used it on the IBM 3033 system to prepare registered (to an ERMAN representation of the Universal Transverse Mercator map projection) landcover classification thematic maps. These were viewed on the colour television display and then unloaded to magnetic tape, where each pixel had its classification status recorded as a decimal number. These magnetic tapes were then returned to New Zealand where they were reformatted and had their classification status numbers translated into numbers compatible with colours generated on an Optronics C-4300 Colorwrite drum write machine (Burden and Whitcombe 1980, McDonnell 1979). The generated colour transparencies were then converted into print form, perhaps overlaid with topographic linework, and constituted into a registered thematic map. Lineprinter thematic maps, utilising symbols to indicate landcover classes, were also prepared on the IBM Australia Ltd installation through ERMAN. These utilised a special representation of standard lineprinter characters (see Chapter 12) to obviate aspect difficulties in representing pixels on a lineprinter.

Further information on the package is available from:

(i) The *General Information Manual* (IBM 1977) presenting a more detailed overview of the system.

(ii) The *User's Guide* (IBM 1976) is literally that and provided *most* of the answers needed to drive this package.

(iii) An excellent, and readable, theoretical background to the mathematical concepts and terminology that constitute ERMAN may be found in the book *Remote Sensing: The Quantitative Approach* by Swain and Davis (1978).

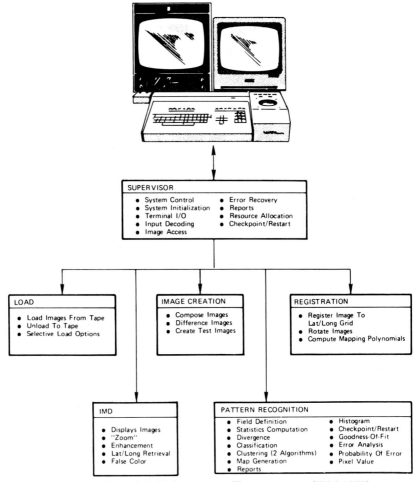

Figure 9.2 Major ERMAN II components (IBM 1977).

9.3 An Analysis Pathway

All operational analysis approaches or pathways are dynamic. They continue to change because users grow in understanding, expectations and involve more concepts and variables as they talk with colleagues. All we can do here is present an aggregated pathway that we followed and suggest it as an initial discussion starter for framing your own first approach. Some of the fundamentals being used—ground truth, accuracies, products—are constants in any pathway.

It is assumed that the user has already marshalled the following: topographic maps at suitable scales, panchromatic air survey photographs, imagery of the database she/he wishes to use as a digital database for the classifier, etc. Similarly it is assumed that the user has decided which classes she/he wishes to be discriminated. Such classes should have been ordered in some hierarchical manner, as outlined in Chapter 3.

The pathway we describe here is an amalgamation of that we followed for the Landsat and Aircraft Multispectral Scanner (AMS) studies reported in Chapters 12–15. One key difference was that the Landsat data had four spectral channels and the AMS had eleven. We ignore the impact of this difference here leaving it to the actual example chapters since we are primarily interested in the general approach common to most systems. Similarly the radiometric/geometric correction of the AMS data, referred to in Chapter 5, is excluded from this presentation.

A typical pathway could involve the following steps.

9.3.1 Gathering of supporting products

The project team leader should have gathered, collated, and marked up the appropriate copies of topographic maps, terrain relief maps, ground water maps, soil maps, forest/land use inventory maps, standard aerial photographs, ground truth inventory returns (needed for training field selection), digital data tapes, photographic products prepared from the digital database (to ensure absence of cloud or its shadow over the target areas) etc. (For special treatment in the processing of these derived photographic products see § 9.3.3.) Likewise some familiarity with the analysis system is needed through system documentation, training courses etc.

9.3.2 Establishment of hierarchical order of classification

As outlined in Chapter 3 it is helpful if the various classes the user wishes to discriminate are ranked according to broad classes, narrower classes and associated subclasses. An appreciation of these relations usually comes easily to discipline-oriented users and assists in focusing thinking on necessary allied and supporting investigations (e.g. assessment of fertiliser applications patterns, spread of disease infestation etc) needed to complement the classification process. Also, if it later proves difficult to discriminate subclasses a rational amalgamation can take place.

9.3.3 Hue and texturally enhanced images

Suitably enhanced photographic products aid the location of ground truth sampling regions for the classification process, and control points for the registration phase. The hue enhanced product presents broad landcover trends. When textural enhancement is combined with hue enhancement, fine detail and departures from 'homogeneity' become more apparent. This also aids the location of ground control points.

9.3.4 Topographic linework overlay

Once the enhanced colour transparency has been prepared, topographic linework held on a transparent foil is easily combined with the image data during the colour printing process. Such superposition is of great value in location work when using the digital data displayed on the interactive monitors.

9.3.5 Ground control point determination

Suitable ground control points should be selected for use in the registration process (see Orti 1978 for comment on the optimal numbers and placing of GCP). These are best located and marked on the texturally enhanced Landsat image and scaled to the nearest second of latitude/longitude from the topographic maps. This level of accuracy was usually found to be adequate for Landsat. If they are evident on the suitably enhanced products it will probably also be possible to locate them on the television display.

9.3.6 Field visit—ground control points and training fields

Once the GCP and training fields have been selected on the basis of the best available data, a field visit by the prospective user to the region under inspection is considered necessary. This is in order to check:

(i) That any ground control points which were difficult to resolve on the image are well documented and photographed to facilitate their subsequent location.
(ii) That the vegetation species for the training field were correct at the time of the imagery. An adjustment for growth, logging practice, or other changes may sometimes be necessary.
(iii) That the vegetation species over the training field is indeed homogeneous. If it is not, the boundary of the 'field' can be modified or replaced by a more homogeneous field.

9.3.7 SYSTEM step 1: synthesis of extra bands

From the four Landsat bands the twelve ratioed bands and four Principal Component bands can be synthesised, giving a database totalling twenty channels for a single data acquisition by Landsat. (Refer to Chapter 6 for a discussion on Band Ratios and Principal Components and their 'place' in classification.)

9.3.8 SYSTEM *step 2: input of training fields*

The appropriate system segment (see figure 9.1 for ERMAN) is entered and the user outlines the required training fields. The preparatory work in the previous non-SYSTEM steps supports this phase. This is accomplished by outlining the field with a number of corner points, their position being indicated on the colour display monitor via a trackball cursor or similar system.

9.3.9 SYSTEM *step 3: statistics compilation*

The means, standard deviations and variance/covariance matrices over all channels are now computed for all classes. As the Maximum Likelihood method of classification (Swain and Davis 1978) was employed for the New Zealand work, the probability distributions were based on these parameters. Normal distributions for the spectral signature statistics for each class were assumed, hence there was a need to choose homogeneous training fields.

9.3.10 SYSTEM *step 4: statistics evaluation*

From the means and standard deviations prepared under § 9.3.9, an assessment of class separability and training class homogeneity may be made.

Firstly any training classes whose standard deviations appear to be large should be marked for reassessment. A comparison then, within that class, of the means and standard deviations on an individual field basis, can sometimes indicate the presence of subclasses other than those desired. Also if the standard deviation in a given channel for a field, against those of other fields in that class, appears abnormally high, it may point towards a data 'drop-out' or calibration problem in that channel for the area covered by the field.

Secondly an inspection of the class means and their attendant standard deviations may indicate classes that have been badly defined or are actually located closer together in multichannel feature space than originally thought. A comparison with the hierarchical plot for the required classes will assist in resolving (or re-ordering) any interclass distinctions. It is also possible that a class has changed from its original description in the ground truth returns. For example: a farmer may have indicated a field to have had 'barley' in it. However, at the time of sensor recording it may actually have been cut making it appear closer to 'grazed dry pasture'.

9.3.11 SYSTEM *step 5: classification of training fields using the basic data channels*

A further check on the homogeneity of the training fields can now be employed. The training fields alone are classified using the basic data channels. Any fields whose percentage of correct classification declines below 66.7% (say) (± 1 standard deviation in normal statistics) can be re-examined for overall homogeneity or edge effects. A threshold of 0% is used in this reporting phase so that all the pixels in each training field are included.

The steps in §§ 9.3.7–9.3.11 can be repeated until satisfactorily 'pure' training fields have been defined for the system.

9.3.12 SYSTEM *step 6: channel selection via a Separability Index*

As discussed earlier (Chapter 7) a variety of Separability Indices can be used. In the ERMAN system the Divergence function was used and the technique/problems associated with this and other functions are outlined in Chapter 7.

The inclusion of the Synthetic Channels can lead to improved species separability. Divergence selects the best channels for *overall* separability or for best separability between a *pair* of classes. Channels that are less than optimum can be excluded from classification if the channel spectral distributions are non-normal. This improves separability, as well as reducing classification computer time.

One of the constraints on the Maximum Likelihood classifier is that there must be ($n + 1$) different spectral signature points (pixels) in the training data for each class to set up the statistics for a probability distribution over n channels. If this is not possible it is likely that the variance/covariance matrix will prove unable to be inverted. Divergence can be used to reduce the number of channels that enter the Classification step to those that are most effective—if the number of data points does approach the number of channels (as can happen in a practical user case).

The Divergence module as set up in ERMAN operated to select the best 1, 2, . . . , 8 channels from the full set of channels specified. The number of calculations involved, see Chapter 7, precludes more than eight channels at a time. Management of the module, as outlined here, permits the best set, in excess of eight, to be selected.

The channel sets (of eight channels/set) are ranked according to the average value of Divergence over all class pairs. Any class pair that has 'poor' separability, arbitrarily taken as a class pair Divergence of less than 900, for the first ranked channel set of eight, is noted for further evaluation. The best eight channels ranked by the average Divergence, over all class pairs, are selected.

Those class pairs with 'poor' separability are now passed through

Divergence for all permitted channels and the channel sets ordered according to the interclass pair Divergence values. The best channel set is then noted for each of the previously 'poor' class pairs and these channels combined with those ranked first by the average Divergence value, as outlined above, into a composite set for classification.

Similarly by monitoring interclass separability as a function of added channels a plateau (see figure 7.8) can be defined and a compromise between added channels and increased computer time reached.

9.3.13 SYSTEM *step 7: classification of training fields using the Separability Index recommended channel set*

As a further check on the suitability of the training fields, they can be reclassified using the recommended channel set from § 9.3.12. The percentages of correct classification of each of the classes, based solely on the training fields, should hopefully now increase. If not, a return to § 9.3.8 should be contemplated. Again a 0% threshold should be used in the reporting phase since we are concentrating only on the training fields.

9.3.14 SYSTEM *step 8: classification of the desired region*

The full area required to be classified is now classified and reported. However a threshold of 1%, at least, must now be used. This imposes a realistic boundary upon pixels which have low class likelihoods and which are included in classes other than those explicitly specified for this classification. However it does rely upon the initial class statistics having only small standard deviations. These will then be consistent with the homogeneous training fields which have resulted in unimodal statistics.

9.3.15 SYSTEM *step 9: Registration*

The classified image may now be registered onto the desired map projection (see Chapter 8) using 80 m, or more preferably, 40 m resampling for Landsat (Berzins 1981) during the Registration process (see IBM 1976 for a description of the ERMAN package options). A nearest neighbour approach must be used in this Registration process (see Chapter 8).

When different basic data sets are to be combined for use in the Classification process, for example when time sequential imagery or other data (e.g. soil map data) are combined with Landsat data etc, it is necessary to perform Registration before Classification. If one acquisition of Landsat data is being used alone it is believed that Classification should precede Registration to preserve maximum spectral integrity of the data (Wheeler 1980) (and to reduce computer time). However this does demand that the nearest neighbour approach be used in the subsequent Registration (refer to Chapter 8).

9.3.16 SYSTEM *step 10: preparation of lineprinter symbol maps and colour class maps*

A character lineprinter output product was desired with equal X and Y dimensions. Consequently a special character set was generated and implemented on the IBM 3800 lineprinter in Sydney (Beach 1981). As a result excellent character maps at a scale of 1:37 600 were easily prepared. Each pixel would usually have its allocated symbol for the designated class printed out. Those that had likelihood values below the specified threshold for that class were left blank (see Chapter 12 for an example).

The other output product was the colour-coded class map with six available colours, excluding black and white. It is important not to include too many similar colours into a mapping product as discrimination could become difficult. This difficulty is known as colour aliasing and is seen in the association of a pink, for example, next to a red in one part of an image and being misassociated with a red/pink class. However, in another part of the image the same pink, now adjacent to a blue, may be grouped with a pink/purple class. Careful planning was thus needed to determine which classes should be combined as the same colour and which differentiated. For example, in the pastoral map, all forest classes were mapped as the same colour allowing maximum colour differentiation of the pastoral classes. These colour maps, after viewing, were combined, for the work of Chapters 12–15, into a master multiband classification image and written on tape for subsequent transcription to colour film on the Physics and Engineering Laboratory Colorwrite, in New Zealand. Examples of these products are described in the project Chapters 12–15, and may be found in the colour plates 1–8.

A more complete discussion on the output products and their subsequent treatment to support user needs is given in Chapter 10.

9.3.17 *Field visit—check of classification accuracy*

Following the preparation of the final lineprinter coded classification map (see figure 12.3) with equal X and Y scales, and colour classification prints, a return visit to the field must be undertaken.

A field sampling programme is set up covering all classes representatively over the full region and pixels evaluated as to whether they had been correctly classified or not (see Chapter 11 for details). This forms the basis for confidence, or otherwise, in the final classification product.

9.4 References

Beach D W 1981 Informal communication, IBM, Syndey
Berzins G J 1981 Informal communication, Los Alamos Scientific Laboratory, New Mexico

Burden A K and Whitcombe A N R 1980 *FORMAT—a Program Used to Edit the ERMAN Series of Magnetic Tapes on the Varian Computer* Report No 669 (Lower Hutt, NZ: Physics and Engineering Laboratory, DSIR)

IBM 1976 *Earth Resources Management II (ER-MAN II) User's Guide* Program No 5790-ARB Doc. No SGB 11-5008-0 (Brussels: IBM)

IBM 1977 *Earth Resources Management II (ER-MAN II) General Information Manual* Program No 5790-ARB Doc. No G B 11-5007-0 (New York: IBM)

McDonnell M J 1979 Computer image processing and production at PEL *Proc. 10th New Zealand Geography Congress and 49th ANZAAS Congress Geographical Sciences, Auckland* (NZ: NZ Geographical Society) p 274

Orti F 1978 *Optimal Spatial Distribution of Ground Control Points to Minimize Mean Square Registration Error of MSS Landsat Images* Internal Report, IBM Scientific Center, Madrid (see also *Photogrammetric Eng. Remote Sensing* **47** 101)

Swain P H and Davis S M (ed.) 1978 *Remote Sensing: The Quantitative Approach* (New York: McGraw-Hill)

Wheeler S G 1980 Informal Communication, IBM, Houston

10.1 Introduction

Classification data must be presented with visual appeal and in such a way that the major patterns are clearly apparent. To minimise the importance of data presentation is to waste the resources used to produce the classification results.

The type of final classification product required by a user depends on

(i) who the users will be
(ii) how the results are to be displayed
(iii) the degree of detail required
(iv) the scale of the final product.

The product could range from a detailed character map for inventory work, to a generalised colour classification map at a scale of 1:1 000 000 for policy making. Alternatively, the classification results may be used directly for mapping.

Each of these products will be discussed in the following sections.

10.2 Colour Classification Images

Eight colours, including black and white, may be considered as a useful compromise between colour aliasing (see § 9.3.16) and the number of classes to be portrayed. The eight colours suggested here are: black, blue, green, cyan, red, magenta, yellow, white. The technique to handle larger numbers of classes is to refer, probably, back to the classification class hierarchy table and choose one colour for each of the major groups of classes classified, i.e. green for all non-wheat classes, and the remaining colours excluding black and white for the wheat subclasses. Traditionally black can be used to display areas that have either been left unclassified or have been thresholded out in the Maximum Likelihood case.

As an example of this, the breakdown of the colour images for the study reported in Chapter 13 is presented in table 10.1(*a*) with the key

Table 10.1(a) Colour allocations for each class for nine classification images used in support of the Land Cover study (Chapter 13).

Image theme	No threshold	Agricultural	Undeveloped	Bush scrub	Water	Forest	High vigour pasture	Woodlands	Generalised
Reverting	10	10	10	14	14	14	10	10	10
High vigour	4	4	14	14	14	14	4	4	8
Urban	6	6	6	14	6	14	14	6	6
Bare ground	6	6	6	14	6	14	14	6	6
Developed	8	8	14	14	14	14	8	4	8
Undeveloped	12	12	8	14	14	14	12	10	10
Wetlands	14	14	14	14	12	14	6	14	2
Water	2	2	2	2	2	2	2	2	2
Scrub	14	14	4	8	14	12	14	12	12
Fern	14	14	12	8	14	12	10	12	12
Exotic	14	14	14	4	14	10	14	8	4
Forest	14	14	14	4	14	6	14	14	4
Sunlit	14	14	14	4	14	4	14	14	4
Indigenous	14	14	14	4	14	8	14	14	4
Tussock	14	14	14	12	14	12	14	14	12
Threshold	0	0	0	0	14	0	0	0	0

colour/number table used in ERMAN being presented as table 10.1(*b*). A colour may be allocated to each class, either by the analyst or automatically by the system.

Table 10.1(*b*) Colour to number translation table used in portraying ERMAN classification images.

Colour	Number
Black	0
Blue	2
Green	4
Cyan	6
Red	8
Magenta	10
Yellow	12
White	14

10.3 Black and White Printing Mask Preparation

A positive mask of each class can be prepared on which the one class is displayed as black, and all other classes as white. These masks may be produced as individual transparencies and then combined to produce a colour map via the lithographic process.

10.4 Character Maps

Each class is allocated a symbol and the character maps generated by a classification map option may be printed out. For the New Zealand study these were printed at a scale of approximately 1:37 800 on the IBM 3800 laser lineprinter. Each character represented an $80 \times 80 \text{ m}^2$ area on the ground for the Landsat registered datasets—see Chapters 12 and 13 and figure 12.3.

10.5 The Classification Dataset

Once a classification has been performed using a Maximum Likelihood classifier it is usual to produce a classification dataset on disc. This contains, for every pixel in every line, a value that points to a character in a class/character allocation table. This represents the most preferred class for that pixel and a second 8-bit byte (say) contains the likelihood value (on a scale 0 to 255) associated with that preferred classification. The status of each pixel is thus represented as a concatenation of two 8-bit bytes.

This classification dataset is usually 'filtered' by the introduction of a suitable threshold. The threshold represents the cut-off level, below which pixels have likelihood values that are unacceptable and are thus regarded as being 'threshold out'—or less precisely being members of the 'unclassified' class (see Chapter 7).

The classification dataset forms the basis for the colour-coded images which may be unloaded to tape as well as the character maps printed out on the system lineprinter.

It can also be possible to unload the classification dataset itself from disc to tape. With some system tailoring it is therefore possible to prepare an output data tape of this classification dataset for use on other machines.

10.6 Colour-Coded Images to Photographic Prints

Each of the images created on such a system was able to be produced in transparency form as described previously (see § 9.2.1.1). These transparencies were then able to be photographically enlarged and combined with linework by a photographic processing and printing agency.

10.7 Linework Overlays

The registration of classification data to a mapping projection similar to conventional map products allows cartographic linework to be combined with the classified data. Unless a user is extremely familiar with an area, location is difficult without this overlay.

The information necessary to provide a useful overlay is a combination of the cartographic road, river topographic and typographic detail. These sheets are photographically combined onto clear film and a negative prepared for combination with the classification data. White linework over the colour classification image and black over the character-coded printouts proved to be the most legible.

Examples of these products are presented in Chapters 12–15 and the colour plates.

10.8 Low Cost Composite Maps Prepared from the Lineprinter

The lineprinter character maps can be joined together and then cut into areas which correspond to conventional map sheets. For details refer to Chapters 13 and 14. Cartographic detail as described in § 10.7 may be enlarged to the character map scale and photographed onto clear film.

Road and river courses are then traced over a light table onto the character map sheets. These products are ideal to assist in the field checking of a classification.

Such a character map is able to display all the classes of the classification on one product.

Both the conventional linework and the corresponding character map sheets can be enlarged/reduced by photocopying to a suitable scale and the information combined. The result is then printed on an opaque film which can then be used to produce inexpensive prints as required, as presented in Chapter 12.

11 Determining the Confidence Level for a Classification

11.1 Introduction†

The allocation of a confidence level to a classification product is considered essential. The acquisition of site-specific data to check the classification is discussed. A statistical approach to the determination of an appropriate confidence level from these check data is presented. Allowance for human assessment and counting errors is included. The approach is directed towards the discipline-oriented user of remote sensing data and is illustrated with actual test data.

The emphasis here is on the use of a Maximum Likelihood classification system but the principle behind deriving a meaningful confidence level may be extended to apply to all classification approaches be they parametric or non-parametric (as outlined in Chapter 7).

11.2 Classification Methodology

A classification is regarded as consisting of the following components:

(i) the acquisition of data;
(ii) a decision by the user as to the level of class separability that is desired and can be attained, being mindful of the spatial averaging of ground cover classes produced by the data acquisition sampling system;
(iii) the selection of training areas which will suit a given computer-based classification software package;
(iv) the effective operation of the package;
(v) the selection of an appropriate threshold for each class to apply to the likelihood distribution for that class;
(vi) the creation of an appropriate output product after the thresholds have been applied.

† The original of this chapter appeared as an article under the above heading by I L Thomas and G McK Allcock (1984 *Photogrammetric Eng. Remote Sensing* **50** 1491–6) and is included here with permission.

For this illustration we consider Landsat to be the data acquisition system and the IBM Earth Resources MANagement package (ERMAN) as the analysis software (IBM 1976).

As we've discussed previously, a data acquisition system produces a set of numbers for each spatial resolution element, or pixel, on the assumption that each element is homogeneous. However, generally the ground cover within a given element will in fact be heterogeneous to some degree. The analyst must, therefore, decide whether the assumption of homogeneity is acceptable in each case.

Analyst interaction with a software package inserts appropriate training area characteristics into the classification process. Effective operation of the package requires adequate expertise on the part of the operator as well as accurate software, sufficient mathematical and scientific precision in the technique, and appropriate delineation of class boundaries within the training data.

Each pixel is classified, using ERMAN, into the most likely class type and has an appropriate likelihood associated with it (Swain and Davis 1978, Chapter 7). Obviously, if the number of classes chosen for the classification process does not include all the classes of the area being classified, some pixels will be given an incorrect class association although, hopefully, with a low likelihood. Each class can, however, have associated with it a minimum threshold above which pixels may confidently be expected to be members of that class. The assignment of a specific minimum threshold to each class is a decision that must be made by the analyst.

The creation of an appropriate output product can also include a decision by the analyst. If the computer-produced data are further processed or interpreted to prepare a land use map, then allowance must be made for the impact of further human decisions.

Consequently, the success of a classification can be influenced by a variety of factors—sensor, software, and human. The derivation of a confidence level for a classification must recognise that it represents such a combination of influences.

11.3 Why have a Confidence Level?

Any computer-derived classification that will lead ultimately to a ground cover thematic map is based on ground truth data gathered by the user from selected 'training' areas. This applies whether unsupervised clustering or supervised classification is employed and whether parametric or non-parametric techniques are used.

The computer may represent classes of similar ground cover by character symbols on a lineprinter, coloured picture elements (pixels) on a television monitor screen or numbers on computer tape for subsequent transcription to transparency film.

The accuracy of the thematic map depends on our ability to extrapolate successfully from the training areas to the whole mapped area. Unless we have some statistical measure of the efficiency of the extrapolation process, we cannot estimate our level of confidence in the classification. Once a confidence level is so quantified, then a user of the classification data can relate it, via the probability of correct classification, to actuality over the whole classified area. The classification is thus married to ground actuality by the confidence level. (Here we are using the term 'ground actuality' to distinguish, and stress, the difference between the set of check data used to evaluate the classification product and the 'ground truth' data used to set up the statistics. Hence the form of the classification can be ascertained. The two sets of data must obviously be separate but they must equally have the same characteristics of location, type, height, health, etc. Often the ground truth data are taken from a well controlled and known area whereas the ground actuality data are taken over a wider area, over less pure ground cover pixels and employs essentially random sampling. Thus, the ground actuality data more closely represent what is actually covering the ground whereas the ground truth data are usually aimed at the purest classes to effect best class separations in the classification process.)

11.4 What is a Confidence Level?

A pixel classified into a particular class can only be correctly or incorrectly classified. There is no middle ground. That is, if the probability of correct classification of a pixel belonging to a given class is p, and the probability of incorrect classification is q, then

$$p + q = 1. \tag{11.1}$$

In this case, for one pixel taken at random from the complete set of pixels belonging to the class, the probability P that the pixel is correctly classified is $P(1) = p$. Similarly, the probability of being incorrect is $P(0) = q$.

For a two-pixel sample from the complete set there are four possible combinations, where R indicates the classification has been shown to be correct and W indicates an incorrect result:

$$RR, \ WR, \ RW, \ WW.$$

Here, the probability of being correct twice $P(2) = p^2$; the probability of being incorrect twice is $P(0) = q^2$; and the probability of having one correct and one incorrect is $P(1) = 2pq$ (we are not concerned with sequential ordering). Similarly, for a three-pixel sample we could have the combinations

$$RRR, WRR, RWR, RRW, WWR, RWW, WRW, WWW.$$

The probabilities then would be

$$P(3) = p^3 \qquad P(2) = 3qp^2 \qquad P(1) = 3q^2p \qquad P(0) = q^3.$$

Translating to numerical probabilities, if $p = \frac{3}{4}$ and $q = \frac{1}{4}$

$$P(3) = 27/64 \qquad P(2) = 27/64 \quad P(1) = 9/64 \quad P(0) = 1/64.$$

The development of a probability distribution can be noted in the above examples, where the abscissa represents a stipulated number i of correctly classified pixels in a sample of n pixels, and the ordinate represents the probability that the number of correctly classified pixels is found upon examination to be exactly equal to i.

Three other points also emerge:

(i) The probabilities to be associated with 0, 1, 2, . . ., n correctly classified pixels from a sample of n pixels drawn from the complete set may be given by the terms in the binomial expansion of $(p + q)^n$.

(ii) As $p + q = 1$ then $(p + q)^n = 1$, hence

$$P(n) + \ldots P(3) + P(2) + P(1) + P(0) = 1.$$

(iii) The coefficient for $P(i)$ is given by the Binomial Coefficient C_i^n where

$$C_i^n = \frac{n!}{i!(n-i)!}. \tag{11.2}$$

Consequently, we may represent the probability of finding exactly i pixels correctly classified in an n-pixel sample as $P(i)$ where

$$P(i) = C_i^n p^i q^{n-i}. \tag{11.3}$$

This obviously leads to a distribution of $P(i)$. It is known as the Binomial Distribution. The mean (m) and standard deviation (s) for the Binomial Distribution† are (Moroney 1956, p 124)

$$m = np \tag{11.4}$$

$$s = (npq)^{1/2}.$$

---

† Strictly speaking, the Binomial Distribution requires that each time we examine whether or not a randomly selected pixel is correctly classified, we should immediately replace the pixel, so that we are always selecting from the complete set (consequently with a finite chance that the same pixel is selected more than once). In practice, we sample *without* replacement, in which case the Hypergeometric Distribution should be used (Aitken 1942, pp 56–8). However, provided that we are dealing with large sample sizes, the properties of the two types of distribution can be assumed to be identical.

Usually we are concerned, when checking the efficiency of a classification, with the summation of the probabilities for all stipulated pixels between n and a lower bound, say i. That is, we wish to know the probability that at least i pixels are correctly classified, when a random sample of n pixels is selected.

This probability is called the Confidence Level (CL) for that classification, and is usually expressed as a percentage. Thus, if CL is the integrated probability expressed as a percentage, we can say that we are CL% confident that the pixels are classified correctly at least i times out of n (or at least $(100\ i/n)\%$ of the time).

The mathematical evaluation of CL from the Binomial Distribution rapidly becomes tedious. A more convenient approach is sought.

Mood (1950, p 139) demonstrates that as the sample size n becomes larger the discrete Binomial Distribution approaches the continuous Normal Distribution, as the limiting case for n tending towards infinity. If the *total population* has both a finite mean and standard deviation, then the *sample* mean and standard deviation may be described by equation (11.4), again for an increasing sample size (Mood 1950, Moroney 1956). (This is based on the Central Limit Theorem and applies without reference to the population distribution function form, provided that large samples are involved. By large, a sample of fifty should be regarded as a minimum (Unthank 1960) with a sample in the hundreds (300, Mood 1950) being more acceptable.) These conditions would usually be met by the practical classification tasks we are addressing here.

Consequently, under these conditions, we use the more mathematically tractable Normal Distribution. This is especially useful as when the total area under the curve is normalised to 1.0 the probability we seek is the integrated area between the limits appropriate to n and i. The equation for this unit-area Normal Distribution is (from Moroney 1956, p 117)

$$\text{Probability density} = \frac{1}{s(2\pi)^{1/2}} \exp\left(\frac{-(i - m)^2}{2s^2}\right). \qquad (11.5)$$

Van Genderen *et al* (1978) show that the number of samples necessary to support the achievement of the desired confidence level in the classification product is a function of that required level. For example, for the attainment of the Anderson *et al* (1972) suggested level of 90% confidence in a classification, Van Genderen *et al* (1978) conclude that 30 *randomly distributed* samples are necessary *as a minimum* to support such an assessment. This is discussed further by Rosenfield *et al* (1982). (Compare back to the sample sizes felt necessary to permit the Binomial Distribution to be replaced by the Normal Distribution.)

The task is now to redefine the CL in terms of equation (11.5). Under such a curve the integrated area from three standard deviations below the mean to plus infinity is 0.999. Thus, if we wish to have 99.9% confidence in our evaluation of the performance of the classifer, then the lower bound to the number of pixels that must be correctly classified in the check sample is equal to the mean minus three standard deviations.

11.5 Ground Actuality Checking of the Classification

Obviously, it is impossible to check every pixel of a classified area. By taking a suitably selected sample of pixels, representative of all conditions of vegetation/soil/climate etc that exist over the area, statistical techniques can then lead to a representative confidence level for the classification. Van Genderen *et al* (1978) outline factors that should be borne in mind when designing any sampling programme. They further indicate a simple and acceptable method for establishing a network of sampling sites. This is to support the checking of the required number of samples for the desired level of classification confidence. (However, they do point out that limitations to access may intrude upon the physical implementation of such a sampling programme. This, as indicated later, did modify the sampling programme used to acquire the test data reported here.)

There is no substitute for field checking the classified dataset against the actual conditions that prevail pixel by pixel on the ground. This is known as the site-specific approach (Mead and Szajgin 1982).

An alternative, that of checking other classifications or prepared maps, involves another set of human decisions in the process and can only degrade the checking process. Another alternative is to check a multispectral classification using panchromatic air photographs. This also reduces the amount of reliable information that can be applied to the checking process.

Landsat, or any such sampling system, inevitably impresses a sampling grid over the varying ground cover. Allowance must be made for the positioning of this spatial sampling grid when checking the classification against the actual ground cover. The representative ground cover class for each pixel, or sampling unit, must be determined and used. If the class resolution so imposed is not detailed enough then a different sampling system, for example an aircraft scanner, should be employed.

The approach used by the New Zealand group is to take site-specific ground truth, distributed over a wide area, by actual on-site inspection. This covers the geographic extent of the classification and includes representative data on different soil types, local area climatic conditions (e.g. slope aspect, swamp/arid conditions, exposure to wind, sun, etc) and differing cropping cycles, etc. The classification is then set up in a supervised manner by using *part* of the ground truth to provide the training areas. The *remainder* of the ground truth can then be used to check the classification accuracy outside of the training areas. A variation of this technique is also used for those areas that have long-lived ground cover. Here the classification result is taken into the field in lineprinter format as indicated in § 10.8. Individual pixels are checked and marked off for accuracy of classification by on-site comparison. The lineprinter product is ideal for this application as it more easily permits pixel location and recording than the photographic products.

Returning to the degradation in spatial resolution occasioned by the sampling technique: it is obvious that allowance must be made for this in checking a classification product. Field checking therefore must be restricted to those areas separated from the road edge or any other similar clearly non-homogeneous ground cover classes by at least one and preferably two pixels.

If an influence from soil or microclimate is suspected, a subdivision of the check statistics into appropriate soil/microclimate regions is necessary. The computation of individual confidence levels for each class within each of these regions and a comparison of the results then aids an assessment of the influence level of these factors.

The sampling must also be as representative as possible of the whole classified area. A random distribution of such sampling points over the whole region must be sought (Van Genderen *et al* 1978).

The above were the ground rules used by the New Zealand group when checking their ERMAN computer classification results.

11.6 Determination of a Confidence Level

If

N is the number of samples taken;
P is the number of samples that have been correctly classified;
Q is the number of samples that have been incorrectly classified;
m is the (estimated) mean of the distribution;
s is the (estimated) standard deviation of the distribution;
e_m is the standard error of the estimate of the mean;
e_s is the standard error of the estimate of the standard deviation;
e_p is the experimental (human) error in assessing and counting the number of samples that are correctly classified

and

$$p = P/N \qquad\qquad (11.6)$$

$$q = Q/N \qquad\qquad (11.7)$$

then

$$p + q = 1 \tag{11.1}$$

$$m = Np \qquad s = (Npq)^{1/2} \tag{11.4}$$

$$e_m = \frac{s}{\sqrt{N}} \qquad \text{(from Moroney 1956, p 137)} \tag{11.8}$$

$$e_s = \frac{s}{\sqrt{2N}} \qquad \text{(from Moroney 1956, p 137).} \tag{11.9}$$

It is assumed that p is greater than 0.1 and that N is greater than 50, so that the Binomial Distribution may be adequately represented by the unit-area Normal Distribution (Moroney 1956, p 128).

The error e_p is regarded purely as a human assessment and counting error. 'Assessment', in the sense that a field check of a microscopically *heterogeneous* ground cover pixel must produce a dominant class which is regarded as describing that pixel *homogeneously*. This is a human decision. Similarly, counting techniques will have a human error associated with them. The 'assessment' error is minimised by having the same person who set up the ground truth files, trained the computer classification software and selected the thresholds, also doing the ground checking over the whole area. An accompanying impartial observer can also be used to assist in resolving the 'yes/no' status of any dubious pixel classifications during the checking process. Errors in ground cover class interpretation can thus be reduced. This was done by the New Zealand group. Consequently, it was felt that the 'assessment' error would be absorbed into the overall classification error, under these conditions. The counting inaccuracy in a test case involving some 25 000 pixels was found to be less than 0.5%. This was determined by repeated checks of the same data by different analysts with at least two sets of counts per analyst. e_p was then taken (another human decision) to be 0.5%.

The Normal Distribution, normalised to unit area, allows us to determine the number of correctly classified pixels necessary to maintain a confidence level of 99.9%. This is equivalent to determining the lower acceptable limit for a number of correctly classified pixels as being at the mean of the population distribution minus three standard deviations (see § 11.4).

In practice, values for both the mean and the standard deviation are obtained from a restricted (though possibly large) sample drawn from the whole population. Hence these are *estimates*, whose standard errors are given by equations (11.8) and (11.9). These equations indicate that e_m and e_s differ from s by numerical factors only, indicating complete correlation among the quantities. Thus, the value of the lower limit necessary to give a 99.9% confidence level is obtained by taking a value for the mean which is three standard errors *lower* than the estimated mean, and then subtracting three times a standard deviation which is three standard errors *greater* than the estimated standard deviation. That is

$$99.9\% \text{ CL} = (m - 3e_m) - 3(s + 3e_s). \qquad (11.10)$$

Examination of equations (11.4), (11.8) and (11.9) will readily show that when m and N are very large, as in the example below, the standard errors are trivial and can be neglected, in which case

$$99.9\% \text{ CL} = m - 3s. \qquad (11.10')$$

Nevertheless, for the sake of completeness, we have included e_m and e_s in the following illustrative calculations.

As an example of the above approach, we take an actual case of classifying 145.4 km × 117.1 km (2 661 552 Landsat pixels) of the King Country region using the ERMAN package (this is taken further in Chapter 13).

Because the King Country has highly dissected topography it was not possible to access at ground level such a random sampling network as suggested by Van Genderen *et al* (1978). Consequently the site-specific field checking was conducted by driving over most of the road network that existed in the classified land cover area. The pixels were evaluated at least one pixel away from the road edge and along the adjacent ridge lines. The road network spanned the complete area classified and was believed to thus fulfil reasonably well the random sampling criterion.

25 773 pixels were field checked	$(= N)$
(0.97% of the total area)	
24 587 were found to be correctly classified	$(= P)$
1 186 were found to be incorrectly classified	$(= Q)$.

In this example we take the complete classification, not by class, and assess an overall probability for the full classification. Therefore from

$$N = 25\,773 \qquad P = 24\,587 \qquad Q = 1186$$

then

$$p = 0.9540 \qquad q = 0.0460 \qquad m = 24\,587$$
$$s = 33.637 \qquad e_m = 0.210 \qquad e_s = 0.148.$$

From equation (11.10) the lower acceptable limit to give a 99.9% confidence level is:

$$(m - 3e_m) - 3(s + 3e_s)$$
$$= 24\,484 = 95.00\% \text{ of the sample.}$$

However, the above result does not take account of the counting error e_p, which we have earlier set at 0.5%. Thus, 129 pixels (0.5% of 25 773) may have been miscounted. Therefore, to maintain our 99.9% confidence in the result, we must reduce the lower acceptable limit by 129, i.e. to 24 355 = 94.50% of the sample.

We therefore conclude, with 99.9% confidence, that at least 94.50% of the pixels in the *whole* area have been correctly classified. That is, if 1 000 random samples, each of about 25 000 pixels, were taken from the whole 2.66 million pixels being classified, in only one case would we expect to find a classification accuracy of less than 94.50%.

Lesser degrees of confidence may be acceptable in some applications. For instance, if a confidence level of 99% was required, all the 'threes' in equation (11.10) would be replaced by 2.33. To achieve 95% confidence, the 'threes' in the equation would be replaced by 1.65. Applied to the present example, after taking account of e_p as indicated above, the following results are obtained.

$$\text{We are} \begin{Bmatrix} 99.9\% \\ 99\% \\ 95\% \end{Bmatrix} \text{confident that at least} \begin{Bmatrix} 94.50\% \\ 94.59\% \dots \\ 94.68\% \end{Bmatrix}$$

. . . of the pixels in the whole area have been correctly classified.

The very small spread in these figures is a direct result of the very large size of the sample taken. This leads to a strongly peaked distribution with a small standard deviation—note that s is only about 0.1% of m.

As already stated, the above probability of at least 94.50% correct classification for a confidence level of 99.9% pertains to the full multiclass classification. A similar evaluation should be undertaken to assess applicable confidence levels for individual classes.

11.7 References

Aitken A C 1942 *Statistical Mathematics* (Edinburgh: Oliver and Boyd)

Anderson J R, Hardy E E and Roach J T 1972 *A Land Use Classification System for Use with Remote Sensor Data* US Geological Survey Circular 671 (Washington, DC: USGS)

IBM 1976 *Earth Resources–Management II (ERMAN II) User's Guide* Program No 5790-ARB, Doc. No SB11-5008-0 (Brussels: IBM)

Mead R A and Szajgin J 1982 Landsat classification accuracy assessment procedures *Photogrammetric Eng. Remote Sensing* **48** 139–41

Mood A McF 1950 *Introduction to the Theory of Statistics* (New York: McGraw-Hill)

Moroney M J 1956 *Facts from Figures* (Middlesex, UK: Penguin)

Rosenfield G H, Fitzpatrick-Lins K and Ling H S 1982 Sampling for thematic map accuracy testing *Photogrammetric Eng. Remote Sensing* **48** 131–7

Swain P H and Davis S M 1978 *Remote Sensing: The Quantitative Approach* (New York: McGraw-Hill)

Unthank E L 1960 *Statistics for Matriculation Mathematics* (Melbourne: Halls)

Van Genderen J L, Lock B F and Vass P A 1978 Remote sensing: statistical testing of thematic map accuracy *Remote Sensing Environ.* **7** 3–14

12 Forest Inventory from Landsat Imagery

12.1 Introduction

The feasibility of using Landsat satellite digital imagery for forest resource management is examined.

Combining available Landsat imagery and computer analysis systems permitted classification maps of forest/scrub classes to be prepared for two regions in New Zealand. These regions represent two distinctly different types of forest and terrain. The central North Island, King Country region is an area of rough, dissected terrain, with mainly indigenous forest, scrub types and rangelands, while the Darfield–Eyrewell region is a highly productive cropping area with some exotic forest plantations on the flat Canterbury Plains.

This chapter describes the steps used in the Landsat/ERMAN digital data analysis: collection of ground truth, preparation of specially enhanced Landsat imagery, registration of the image to a map projection, use of synthetic bands for classification, through to the final classification maps. It details results achieved—classification accuracy (80–90%), registration accuracy (King Country ± 125 m, Darfield–Eyrewell ± 61 m), as well as cost benefits (82%) and manpower savings (93%) between ground and satellite survey methods.

The results show that this method of forest inventory has great potential as a forest management tool.

12.2 The Study Areas

Two project areas, figure 12.1, were chosen to evaluate Landsat imagery: the King Country project region, 16 400 km^2 in the central North Island of New Zealand and the Darfield–Eyrewell project area, 4500 km^2 located west of Christchurch in the South Island.

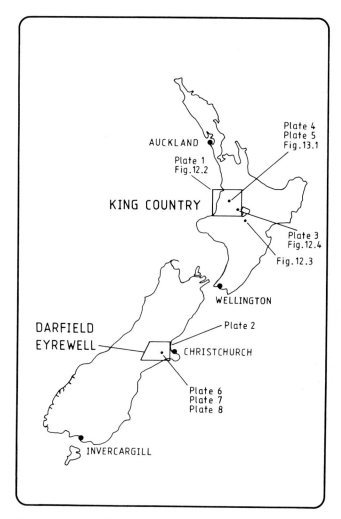

Figure 12.1 Locality map for these studies in New Zealand.

12.2.1 King Country

The King Country area ranges from flat pumice land to steeply dissected mudstones. Dense indigenous forests cover the hills and range and scrublands cover the lower hills. Pockets of high vigour grasslands are found on valley floors and exotic forest plantations on any land cleared of native forest or scrub. Access to the indigenous forest and scrublands is very limited.

The major forest types (Nicholls 1976, Dale 1981) are:

Podocarp forest. A forest type predominantly west of Lake Taupo and in the west King Country, characterised by a tight canopy of mainly rimu (*Dacrydium cupressinum*), matai (*Stachycarpus spicatus*), miro (*Podocarpus ferrugineus*), totara (*Podocarpus totara*), mountain toatoa (*Phyllocladus alpinus*) and kahikatea (*Dacrycarpus dacrydioides*).

Podocarp–tawa forest. The canopy is predominantly tawa (*Beilschmiedia tawa*) with emergent podocarps towering above. This forest type is found in the northern part of the region.

Podocarp–hardwood forest. This contains scattered emergent podocarps, the lower main canopy is kamahi (*Weinmannia racemosa*) with some maire (*Nestegis spp.*), hinau (*Elaeocarpus dentatus*), fuchsia (*Fuchsia excorticata*), broadleaf (*Griselinia littoralis*), tawa and rewarewa (*Knightia excelsa*). This class includes many areas which have been partially logged previously for their podocarp content. Such areas occur widely throughout the region.

Podocarp–hardwood–beech. This is mainly rimu–tawa with irregular mixtures of beech (*Nothofagus truncata, Nothofagus menziesii, Nothofagus fusca*), and is located mainly in the southwest of the region.

Beech. This forest type is virtually pure beech. It is found in the mountain region around Mt Ruapehu and in scattered areas elsewhere.

Exotic plantations. These consist mainly of *Pinus radiata, Pinus contorta, Pseudotsuga menziesii* and some mixed eucalyptus species.

Scrub classes. These are made up of bracken fern (*Pteridium esculentum*), kanuka (*Leptospermum ericoides*), manuka (*Leptospermum scoparium*), exotics (i.e. heath, gorse and broom), broadleaf shrubs, tree ferns, high altitude scrubland and monoao (*Dracophyllum subulatum*).

12.2.2 Darfield–Eyrewell

Darfield–Eyrewell is situated on an alluvial plain and is a highly productive area for cereal crops with some exotic forest plantations on poorer soils and used as wind shelters. The area is well roaded giving very good access into the forests. The main forest species is *Pinus radiata* which ranges in age from new to mature 1950 plantings.

12.3 Spectral Signatures, Landsat Imagery and a Computer Analysis System

All plants reflect differing amounts of red, green, infrared etc light (see Chapter 2).

The reflected or spectral response of plants relates to:

(i) type of leaf (needle, broad leaf)	(v) water content
(ii) leaf cover	(vi) plant condition
(iii) leaf pigments	(vii) plant densities
(iv) leaf cell structure	(viii) leaf orientation.

Training fields of known forest species/mixtures are used to obtain unique reflectance values (spectral signatures) for each class, as outlined in Chapter 9.

Aspect, terrain and microclimates can alter the spectral signatures of a class; therefore it is often necessary to choose subclasses, for example, scrub–sunlight, scrub–shade.

In this evaluation three Landsat images were used:

King Country: scene 2389–21 172 recorded 15 February 1976 (GMT)
Darfield–Eyrewell: scene 2192–21 265 recorded 2 August 1975 (GMT)
                    scene 2282–21 254 recorded 31 October 1975 (GMT).

ERMAN II was the major analysis system for this study. Chapter 9 and the IBM ERMAN *User's Guide* (1977) give in-depth details of the system.

The overall objective of the King Country and Darfield–Eyrewell investigation was to produce a classification map of forest and scrub types using Landsat imagery and a computer-based classification system.

12.4 Project Method

The classification maps were produced using the following steps.

12.4.1 Ground truth collection

Representative areas known as training fields (Joyce 1978) were located for each class. These training fields form the basis for the classification. Identifiable ground control points also had to be located for registering the image.

12.4.2 Hue and texturally enhanced Landsat images

Enhanced Landsat images were prepared which highlighted the colour and fine details associated with forest cover. These images, enlarged to

1:75 000, were used to (i) choose suitable ground control points, (ii) assist in determining the classes that could possibly be discriminated and (iii) locate homogeneous training fields for each class.

12.4.3 Production of supporting map overlays

A road and river pattern overlay was prepared from existing National Mapping (NZMS 242 1:500 000). Although the alignment between the enhanced image and the overlay was not perfect, the overlay could be registered satisfactorily over portions of the image to give location information. The enhanced image was unregistered at this stage and therefore contained the inherent distortions of the Landsat system.

The National Forest Survey of New Zealand, 1955, 1:15 840 scale maps were also produced in overlay form, at 1:75 000, and used in the same manner.

12.4.4 Marshalling of existing forest and land use data

Information was obtained from:

(i)   *The National Forest Survey of New Zealand*, New Zealand Forest Service (1955).

(ii)   Aerial photographs. Survey Nos 2974, 2920, 3838, 5147, 5410, 5014 (1976, 1977, 1979).

(iii)   *King Country Land Use Study: forest and scrub type maps* (1977) (NZMS Series 288, 3 sheets).

(iv)   Exotic plantation record maps, 1:10 000. NZ Forest Service and NZ Forest Products Limited (1979).

(v)   Topographical maps. NZMS 1. 1:63 360.

(vi)   *A revised classification of the North Island Indigenous Forests* (Nicholls 1976).

(vii)   Ecological forest class maps, 1:250 000. NZ Forest Service Mapping Series 6 (FSMS 6) (1979).

(viii)   Geological survey and soil maps, NZ Geological Survey and NZ Soil Bureau (DSIR) (1973).

12.4.5 Field visit 1: familiarisation

A visit was made to obtain an overview of the region and to discuss with local staff the forest types, practices and history. An aircraft was used to view and photograph the forest patterns and terrain. This proved invaluable especially for choosing classes and later during computer processing where quick evaluations were required.

It was also appropriate to involve a system (satellite data and computer processing) specialist in this visit. In our case this supported later enhancement and analysis discussions.

12.4.6 Classes chosen for discrimination

A review of management needs and existing classification systems (Nicholls 1976) was required to establish a list of classes suitable for the forest industry's requirements. Once this had been achieved the enhanced imagery was studied in detail to determine whether a further breakdown of these classes would be required to cope with factors such as: aspect (growth and sun angle), terrain, forest mixtures, plant density, understorey vegetation, plant age, vigour, etc.

29 classes were eventually established for the King Country and 17 classes for the Darfield–Eyrewell region. These are presented as tables 12.1(a) and 12.1(b)

12.4.7 Selection of training fields

Training fields were selected with care to ensure they satisfied the statistical requirements of the Maximum Likelihood classifier and gave a representative sample of the class.

(i)   The *size* of the field had to be large enough to be resolved by the Landsat (1, 2, 3) scanner and thus be located on the Landsat imagery. If possible an area of ten hectares or larger was used. The absolute minimum area was selected as no smaller than four hectares.

(ii)   A *number* of training fields were required for the following reasons: the ERMAN classifier assumes normally distributed data and therefore requires a minimum of $(n + 1)$ pixels per class with sets of radiances that do not match (where n is the number of channels used in classification). It is desirable to have 10–100 times this number to ensure a suitable sample is achieved (Swain and Davis 1978). Because the sample areas, in both regions, tended to be small, it was necessary to select a number of fields that could later be statistically amalgamated to form a suitably sized sample.

(iii)   To obtain a representative sample it was necessary to select training fields with a *uniform and homogenous* forest/scrub cover.

12.4.8 Selection of ground control points

Ground Control Points (GCP) were selected so as to remove inherent distortions of the Landsat data while registering the image to an acceptable map projection. This enabled additional map information such as roads, cadastral boundaries, etc to be overlaid on the final output products and permitted multitemporal analysis of the Darfield–Eyrewell imagery.

To achieve an acceptable accuracy, care was taken in:

(i)   selecting clearly recognisable features on the imagery for GCP;
(ii)   choosing the optimum number and location of points;
(iii) reading the latitude and longitude values for GCP from the NZMS 1 series, 1:63 360, topographical maps.

Table 12.1(*a***)** Classes selected for discrimination in the King Country region.

Class	Description	ERMAN character map symbol used in a final output product (figure 12.3)
Water		W
Bare ground/urban		G
Scrub		
Fern		F
Kanuka/manuka		K
Monoao/tussock		T
Hardwood scrub 1	Sunlit aspect	Y
Hardwood scrub 2	Shady aspect	Z
Alpine/low scrub		A
Broadleaf/smaller hardwoods		O
Indigenous forest		
Tawa	Shaded dense stands	X
Tawa 2	Dense stands	2
Tawa 3	Medium dense stands	3
Tawa/hardwood	Medium-sparse and hardwoods	H
Podocarp	Dense	J
Podocarp/hardwood 1	Medium dense and hardwoods	6
Podocarp/hardwood 2	Medium-sparse and hardwoods	8
Podocarp/hardwood 3	Sparse and hardwoods	9
Beech		L
Kamahi		5
Hardwood		I
Exotic forest		
P radiata† 50	1950 planting	E
P radiata 56	1956 planting	Q
P radiata 62	1962 planting	S
P radiata 70	1970 planting	R
P contorta† 59	1959 planting	D
P contorta 69	1969 planting	C
Ps menziesii† 50	1950 planting	N
Ps menziesii 60	1960 planting	O
Ps menziesii 69	1960 planting	M

Table 12.1(*b*) Classes selected for discrimination in the Darfield–Eyrewell region.

Class	Description	ERMAN character map symbol used in a final output product
Water		J
River gravel		K
Gorse	3 classes, different aspects	6, 7, 8
P radiata 67	1967 planting	R
P radiata 68	1968 planting	S
P radiata 69	1969 planting	T
P radiata 70	1970 planting	U
P radiata 71	1971 planting	V
P radiata 72	1972 planting	W
P radiata 73	1973 planting	X
P radiata 68cc	1968 catch crop planting‡	5
P radiata 69cc	1969 catch crop planting	3
P radiata 70cc	1970 catch crop planting	4
P nigra†		N
Mixed exotics		Q
Forest damage		
Serious windthrow	All trees windblown or broken	P
Moderate windthrow§	Some trees on ground, rest leaning: will not recover	M

† *Pinus radiata* is shortened to *P radiata*; *Pseudotsuga menziesii*, *Ps menziesii*; *Pinus contorta*, *P contorta*; *Pinus nigra*, *P nigra*.

‡ Catch crop, trees destined for shorter life span, under limited silvicultural treatment.

§ Windthrown refers to forest areas damaged in 1975 storm.

GCP were identified on both the enhanced Landsat image and the NZMS 1 maps. Intersections of roads and rivers, bush/road and river/road interfaces and coastal features were the easiest to identify. However, care had to be taken as erosion, bush clearing or even the state of the tide could have altered their positions, especially if the map was a little outdated.

34 points were selected for the King Country and 41 for the Darfield–Eyrewell region. As far as practicable, the points were evenly distributed over the area to be classified. The recommended minimum of 25 points (located in a 5 by 5 pattern) (Chapter 8) was exceeded to allow for redundancies. Figure 12.2 shows the location of the King Country points after the redundant points had been discarded.

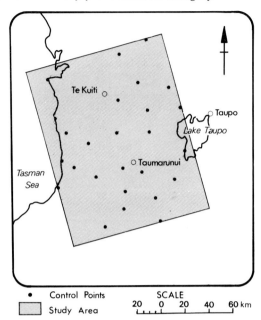

Figure 12.2 Classified area and the location of the ground control points in the King Country region are shown.

The latitude and longitude values of the GCP were scaled to the nearest second of an arc, from the NZMS 1 topographical maps. Because the latitude and longitude graticule formed curved lines the values were determined by proportion. For further details refer to Chapter 8.

12.4.9 Field visit 2: ground control and training field

After initially locating ground control points and training fields, a further field inspection was undertaken.

(i)   All selected training fields were visited to check their homogeneity and class description. Fields were modified, where necessary, either by relocating, or by dividing into other classes. Because of the dense nature of the indigenous forest, it was difficult to inspect training fields from the ground. Along forest road edges, hardwoods tended to flourish and hide the main forest type. Use was made of observation points but these were not always suitably located. A helicopter proved ideal since the training field could be observed at high altitude for location on the enhanced imagery and aerial photographs and checked for homogeneity. Also inspection at 'tree top' level could be used to obtain a description of the main canopy and, if appropriate, the understorey. Exotic and scrub classes could generally be seen from the ground.

(ii) Ground control points could be checked during this visit if problems had occurred in determining their exact loocation.

12.4.10 Computer analysis step 1: Image Registration

Image Registration was undertaken so that the image could be fitted to a map base and to enable the two Darfield–Eyrewell images to be combined for multitemporal analysis. Ground Control Points (GCP), selected in § 12.4.8, were entered by displaying the Landsat image on the colour monitor and indicating each GCP location with the trackball cursor. At the same time the latitude and longitude values were typed in via the alphanumeric monitor. Once all points had been entered, first-order mapping polynomials were calculated and then the X and Y residuals or deviations for each GCP were evaluated. Points with large residuals were discarded and the mapping polynomials recalculated. This procedure was repeated, deleting one GCP each time, until the residuals were less than ± 5 pixels. The King Country image was registered and resampled on an 80×80 m^2 grid, while Darfield–Eyrewell was resampled on a 40×40 m^2 grid. (Refer to IBM 1977 and Chapter 8 for details.)

After studying the results, 40 m sampling would have been preferred for both areas to avoid loss of information, however the four times increase in data would have increased computing times and made data management difficult for the King Country region. This approach is recommended if at all feasible.

12.4.11 Computer analysis step 2: synthesis of extra channels

Four Principal Component channels and twelve Ratioed channels were synthesised from the four registered MSS bands. This gave a total of twenty data channels all registered to the chosen map projection. For a further discussion on these synthesised channels, see Chapters 2 and 6.

12.4.12 Computer analysis step 3: specification of training fields

Training fields that were selected in § 12.4.7 were now entered in the computer analysis system. This was achieved by displaying portions of the image (using the registered MSS bands 4, 5 and 7 as blue, green and red respectively), at magnifications of 7× to 3×, i.e. at scales 1:30 000 to 1:75 000 on the colour monitor. Each training field was then outlined using the trackball cursor.

12.3.13 Computer analysis step 4: compilation of statistics

The Maximum Likelihood classification assumes that the spectral signatures for each class are distributed in a normal Gaussian Distribution. For each class a mean vector and covariance matrix were calculated from the training field data. These parameters were carried into classification.

12.4.14 Computer analysis step 5: classification of training fields using the four MSS channels

By classifying the training field areas only, a further check was made for homogeneity and class separation. Following classification an interclass summary report was prepared. This identified the number of pixels that were classified into each class. If a training field was less than 66.7% correct ($\pm$ one standard deviation in normal statistics), utilising a 0% threshold (since they were simply training fields), then the field was examined and possibly relocated with its boundaries modified, or deleted. The report indicated mixtures which in some cases were accepted. (The training field for podocarp–hardwood 2, for example, had some podocarp and hardwood classes included.)

12.4.15 Computer analysis step 6: channel selection via Divergence

Divergence, chosen here as the Channel Separability approach (see Chapter 7), allowed the user to select the optimum number of channels for classification. This reduced computer time during classification. Separability between classes was gauged from 1 (unseparable) to 999 (easily separable).

The optimum channels were selected by the following procedure:

(i)   The twenty channels and all the classes were introduced to the Divergence module and a request was made for sets of four channels to be ranked according to average Divergence (D_{ave}) (see table 12.2.)

(ii)   Using these results weak class pairs with a value less than 900 were noted (see table 12.3.)

(iii)   Weak class pairs were introduced one at a time to the interclass pair Divergence (D_{ij}) with all channels and again the best four channels were requested to separate these class pairs (see table 12.3 column *a*.)

(iv)   A frequency distribution of the channels recommended in (iii) to separate the weak class pairs was compiled. This enabled the more important channels to be noted and the lesser ones to be left out of the new channel set. In the case of the King Country, those channels that were only required once were left out of the new channel set. The new channel set was compiled from those recommended in (i) and from the frequency distribution of (iii)—based on the concept of additive separability (Chapter 7 and Swain and Davis 1978). The frequency distribution was used to keep channels, and therefore computing time, to a minimum.

(v)   Using the new channel set, classification was entered, the training fields were classified and class summary reports again prepared. The class summary reports were evaluated for unacceptable crosstalk for each class.

For example, if the podocarp–hardwood class summary had some of the pixels classified as podocarp, it was considered acceptable. If some of the pixels were classified as beech within the podocarp–hardwood training areas it was considered unacceptable and these class pairs were noted for further attention (see table 12.3 column *b*).

(vi)   Each class pair with unacceptable crosstalk was then introduced to D_{ij} to again select the best four channels, from the twenty channels.

(vii)   The final channel set was then compiled from those recommended in (iv) and (vi). Again a frequency distribution was used when selecting additional channels from (vi).

Table 12.2 Best four channel sets for King Country forestry study, ranked by D_{ave}. See also discussion on ratioing in Chapter 6.

Set	Channel				Set	Channel			
1	6,	'5/4',	'6/4',	'6/5'	16	PC1,	'4/5',	'6/4',	'6/5'
2	6,	'4/5',	'6/4',	'6/5'	17	6,	4,	'4/6',	'6/5'
3	4,	6,	'6/4',	'6/5'	18	6,	'4/6',	'6/4',	'6/5'
4	PC1,	'5/4',	'6/4',	'6/5'	19	6,	'4/7',	'6/4',	'6/5'
5	6,	'5/4',	'5/7',	'6/5'	20	6,	'4/5',	'4/6',	'6/4'
6	6,	'4/5',	'5/7',	'6/5'	21	4,	6,	'6/5',	'6/7'
7	6,	'4/6',	'5/4',	'6/5'	22	6,	'4/5',	'4/7',	'6/4'
8	6,	'5/7',	'6/4',	'6/5'	23	PC3,	'5/4',	6,	'6/5'
9	6,	'4/7',	'5/4',	'6/5'	24	PC3,	'4/5',	6,	'6/5'
10	6,	'5/7',	'4/5',	'6/4'	25	4,	'4/6',	'6/4',	'6/5'
11	6,	4,	'5/7',	'6/5'	26	6,	'4/6',	'5/7',	'6/5'
12	4,	'5/7',	'6/4',	'6/5'	27	4,	'6/4',	'6/5',	'6/7'
13	6,	'4/5',	'4/6',	'6/5'	28	'4/6',	'5/4',	6,	'6/4'
14	6,	'4/5',	'4/7',	'6/5'	29	4,	'5/6',	6,	'6/4'
15	6,	'5/4',	'5/7',	'6/4'	30	4,	5,	6,	7

	Reference		Reference
PC1		'4/7'	$16 \times \text{MSS}4/(\text{MSS}7 + 1)$
PC2	Principal Components 1, 2, 3, 4	'5/4'	$32 \times \text{MSS}5/(\text{MSS}4 + 1)$
PC3	(Swain and Davis 1978, p 350)	'5/6'	$32 \times \text{MSS}5/(\text{MSS}6 + 1)$
PC4		'5/7'	$16 \times \text{MSS}5/(\text{MSS}7 + 1)$
4	Landsat MSS4	'6/4'	$32 \times \text{MSS}6/(\text{MSS}4 + 1)$
5	Landsat MSS5	'6/5'	$32 \times \text{MSS}6/(\text{MSS}5 + 1)$
6	Landsat MSS6	'6/7'	$16 \times \text{MSS}6/(\text{MSS}7 + 1)$
7	Landsat MSS7	'7/4'	$64 \times \text{MSS}7/(\text{MSS}4 + 1)$
'4/5'	$32 \times \text{MSS}4/(\text{MSS}5 + 1)$	'7/5'	$64 \times \text{MSS}7/(\text{MSS}5 + 1)$
'4/6'	$32 \times \text{MSS}4/(\text{MSS}6 + 1)$	'7/6'	$64 \times \text{MSS}7/(\text{MSS}6 + 1)$

Table 12.3 Divergence results: King Country region.

Using the best recommended four-channel set, ranked by D_{ave} (MSS6, ratios '5/4', '6/4', '6/5') to obtain class pairs with D_{ij} values <999	D_{ij} value	(a) Recommended channels to separate weak class pairs with D_{ij} <900	New D_{ij} value	(b) Recommended channels to eliminate unacceptable crosstalk in training field classification	New D_{ij} value
Fern–Hardwood scrub 1	866	'4/5', '5/4', '7/4', '7/5'	966		
Fern–Tawa 3	997				
Fern–*Ps menziesii* 69	989				
Fern–Hardwood	975				
Kanuka/man.–Monoao/tus.	838	4, 5, '4/7', '5/7'	999		
Tawa–Kamahi	997				
Tawa–*P contorta* 59	977				
Tawa–Podocarp	875	5, '4/5', '5/4', '5/7'	999		
Tawa–*P radiata* 56	997				
Tawa–Tawa 2	784	PC4, '4/5', 4, 5	996		
Tawa–Tawa 3	997				
Tawa–Podocarp	879	'4/6', '5/7', '6/4', '7/5'	999		
Tawa–Tawa/hardwood	988				
Tawa–Kamahi	997				
P radiata 62–*P contorta* 69	875	4, '4/6', '5/4', '5/6'	999		
P radiata 62–*P radiata* 70	997				
P radiata 62–*P contorta* 59	968				
P radiata 62–*P radiata* 56	634	'4/6', '6/4', '7/4', '7/6'	999		
P radiata 62–Tawa 2	986				
P radiata 62–Tawa 3	995				
P radiata 62–*Ps menziesii* 50	929				

Table 12.3 (contd)

Using the best recommended four-channel set, ranked by D_{ave} (MSS6, ratios '5/4', '6/4', '6/5') to obtain class pairs with D_{ij} values <999	D_{ij} value	(a) Recommended channels to separate weak class pairs with D_{ij} <900	New D_{ij} value	(b) Recommended channels to eliminate unacceptable crosstalk in training field classification	New D_{ij} value
P radiata 62–Kamahi	998				
P radiata 62–Broadleaf/sh	998				
P contorta 69–*P radiata* 70	996				
P contorta 69–*P radiata* 56	978				
P contorta 69–Tawa 2	998				
P contorta 69–Tawa 3	998				
P contorta 69–*Ps menziesii* 50	873	4, 5, '4/6', '5/4'	999		
P contorta 69–Kamahi	998				
P radiata 70–Tawa 3	993				
P radiata 70–*Ps menziesii* 69	998				
P radiata 70–*Ps menziesii* 50	837	5, 6, '6/7', '7/5'	968		
P radiata 70–*Ps menziesii* 60	930				
P radiata 70–Kamahi	996				
Hardwood scrub 1–Tawa 3	993				
Hardwood scrub 1–*Ps menziesii* 69	989				
Hardwood scrub 1–Hardwood	991				
P contorta 59–*P radiata* 56	936				
P contorta 59–Tawa 2	996				
Podocarp/hw 1–Beech	995			PC1, '4/5', '5/7', 4	999
Podocarp/hw 1–Tawa 2	928				
Podocarp/hw 1–Podocarp	977				
Podocarp/hw 1–Tawa/hw	992				

Table 12.3 (contd)

Using the best recommended four-channel set, ranked by D_{ave} (MSS6, ratios '5/4', '6/4', '6/5') to obtain class pairs with D_{ij} values <999	D_{ij} value	(a) Recommended channels to separate weak class pairs with D_{ij} <900	New D_{ij} value	(b) Recommended channels to eliminate unacceptable crosstalk in training field classification	New D_{ij} value
Podocarp/hw 1–Podocarp/hw 2	998				
P radiata 56–Tawa 2	995				
P radiata 56–Tawa 3	998				
P radiata 56–Podocarp	997				
P radiata 56–*P radiata* 50	997				
P radiata 56–*Ps menziesii* 50	973				
Tawa 3–Podocarp	997				
Tawa 3–Tawa/hw	976				
Tawa 3–Podocarp/hw 2	993				
Tawa 3–Kamahi	998				
Tawa 3–Broadleaf/shw	963				
Tawa 3–*Ps menziesii* 69	737	4, 5, '4/6', '5/4'	999		
Tawa 3–*Ps menziesii* 50	969				
Tawa 3–*Ps menziesii* 60	993				
Tawa 3–Hardwood	429	(5, 7, '4/5', '7/4'	884)		
		(5, 7, '4/5', '7/4', '7/5'	937)		
		(5, 7, '4/5', '5/4', '7/4', '7/5'	963)		
Tawa 3–Kamahi	963				
Tawa 3–Broadleaf/shw	484	4, 5, '4/5', '5/4'	999		
Podocarp–*P radiata* 50	992				
Podocarp–Tawa/hw	998				

Table 12.3 (contd)

Using the best recommended four-channel set, ranked by D_{ave} (mss6, ratios '5/4', '6/4', '6/5') to obtain class pairs with D_{ij} values <999	D_{ij} value	(a) Recommended channels to separate weak class class pairs with D_{ij} <900	New D_{ij} value	(b) Recommended channels to eliminate unacceptable crosstalk in training field classification	New D_{ij} value
Podocarp–Podocarp/hw 2	998				
Ps menziesii 69–*Ps menziesii* 50	998				
Ps menziesii 69–Hardwood	609	PC2, '5/4', 4, 5	999		
Ps menziesii 69–Kamahi	992				
Ps menziesii 69–Broadleaf/shw	973				
Ps menziesii 50–Kamahi	954				
Ps menziesii 50–Broadlead/shw	985				
Ps menziesii 60–Kamahi	929				
Hardwood–Kamahi	995				
Hardwood–Broadleaf/shw	937				
Tawa/hw–Podocarp/hw 2	976				
Tawa/hw–Broadleaf/shw	998				
Kamahi–Broadleaf/shw	971				
Classes below had unacceptable crosstalk in training field classification					
Tawa–Beech	999			PC1, '4/5', '6/4', 4	999
Tawa 2–Beech	999			PC1, '4/7', 4, 7	999
Tawa 2–*Ps menziesii* 69	999			PC1, '6/5', 5, 6	999
Tawa 2–*Ps menziesii* 60	999			PC1, PC2, '5/7', '7/5'	999

†hw stands for hardwood; shw, smaller hardwoods.

A final channel set of 15 was chosen for the single acquisition Darfield–Eyrewell approach using scene number 2 282–21 254 and a 14-channel set for the King Country scene, number 2 389–21 172.

> Darfield–Eyrewell: MSS 4, 5, 7, Ratios '4/6', '4/7', '5/4', '5/6', '5/7', '6/4', '6/5', '6/7', '7/4', '7/5', Principal Components 2 and 4.
> King Country: MSS 4, 5, 6, 7, Ratios '4/5', '4/6', '5/4', '5/6', '5/7', '6/4', '6/5', '7/4', '7/5' and Principal Component 1. (The algorithms used for the Ratios are given in Chapter 6.)

12.4.16 Computer analysis step 7: classification of training fields using Divergence recommended channel set

A final check was made of the training fields. If at this stage a 90% correct classification, with a 0% threshold, was not achieved, then consideration would have been given to returning to § 12.4.12.

12.4.17 Computer analysis step 8: classification of the desired region

Both the Darfield–Eyrewell and King Country regions were classified using the optimum number of channels, 15 and 14 respectively, for the single acquisition cases.

To enable an evaluation of multitemporal data, a classification of the Darfield–Eyrewell region was also prepared using:

(i)   the four MSS channels of scene 2 282–21 254 and
(ii) the eight MSS channels of scenes 2 282–21 254 and 2 192–21 265.

12.4.18 Computer analysis step 9: preparation of class maps

In using the Maximum Likelihood classifier, every pixel is assigned to its nearest class. Therefore, it is necessary to introduce a threshold to exclude those pixels with a low likelihood of belonging to the class (see also Chapter 7). The threshold value for each class was determined by studying the classification map on the colour monitor and varying the threshold until an acceptable minimum of class overlap was achieved. This approach was particularly important in dealing with mixed species in indigenous forests usually not in conformance to the Landsat sampling interval. From the final threshold values, given in table 12.4, it can be seen that the King Country hardwood scrub 1 and the Darfield–Eyrewell *Pinus radiata* 1971, 1972 and 1973 all have high threshold values. This can be attributed to a wider statistical distribution than in other classes. Hardwood scrub 1 covered many types and densities of scrubs while the *Pinus radiata* classes had a large component of bare ground/understorey scrub since the forest at that stage had an open canopy. The 0% thresholds were attributed to the close association of classes and therefore the closeness and partial overlapping of statistical distributions (Swain and Davis 1978).

Character classification maps (see figure 12.3) were prepared for each region using the threshold values in tables 12.4 and at a uniform 0% threshold value. These maps were prepared using modified software (Beach 1980) on the IBM 3800 laser lineprinter to give equal X and Y scales at 1:37 800.

Table 12.4(*a*) Thresholds used for the King Country study.

Class		Threshold (%)	ERMAN character map symbol (In final output product figure 12.3)
Water		0	W
Bare ground/urban		1	G
Scrub			
Fern		0	F
Kanuka/manuka		0	K
Monoao/tussock		0	T
Hardwood scrub 1		33	Y
Hardwood scrub 2		1	Z
Alpine/low scrub		0	A
Broadleaf/smaller hardwoods		0	O
Indigenous forest			
Tawa		0	X
Tawa 2		0	2
Tawa 3		0	3
Tawa/hardwood		0	H
Podocarp		0	J
Podocarp/hardwood 1		0	6
Podocarp/hardwood 2		0	8
Podocarp/hardwood 3		0	9
Beech		0	L
Kamahi		0	5
Hardwood		1	I
Exotic forest			
P radiata	50	1	E
	56	1	Q
	62	1	S
	70	1	R
P contorta	59	1	D
	69	1	C
Ps menziesii	50	1	N
	60	1	O
	69	1	M

Table 12.4(*b*) Thresholds used for the Darfield–Eyrewell studies.

Class		8 MSS channels	4 MSS channels	15 optimum channels	ERMAN character map symbol (in final output product)
		Threshold (%)			
Water		1	1	1	J
River gravel		1	1	1	K
Gorse		1	1	1	6, 7, 8
P radiata	1967	1	1	1	R
	1968	1	1	1	S
	1969	1	1	1	T
	1970	1	1	1	U
	1971	30	60	60	V
	1972	30	40	40	W
	1973	1	30	30	X
	1968 CC	1	1	1	5
	1969 CC	1	1	1	3
	1970 CC	1	1	1	4
P nigra		1	1	1	N
Mixed exotics		1	1	1	Q
Forest damage					
Serious windthrow		1	1	1	P
Moderate windthrow		1	1	1	M

ERMAN had six colours, plus black and white, available for displaying classification maps. Careful planning had to be undertaken to ensure the various facets of forestry were shown since there were 16 classes in Darfield–Eyrewell and 29 classes in the King Country. This entailed the combining of classes into the same colour, e.g. the King Country scrub map shows the indigenous and exotic forests as one colour, and the remaining colours were used to differentiate the scrub classes.

Colour maps were unloaded onto magnetic tapes for subsequent transcription to photograph transparency material through a drum writing machine (an Optronics C-4300 Colorwrite). The various options for the output products are detailed further in Chapter 10.

12.4.19 *Computer analysis step 10: preparation of character map overlays*

To assist with field checking the classification, a cartographic overlay was prepared by enlarging existing NZMS 1 maps to the character map scale.

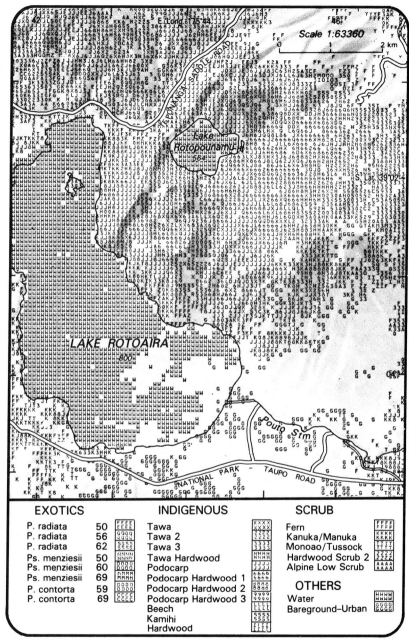

Figure 12.3 Character-coded classification map prepared from the classification data set using the threshold values given in table 12.4(*a*) being output onto a lineprinter. Topographic data were then photographically overlayed to produce a final product that assisted in the location of ground cover classes classified from the Landsat image, scene reference number 2389–21172.

The relationships between the latitude/longitude and the corresponding pixel row/column values were first obtained for a number of points throughout the image. These points, chosen to fall in the four corners of the NZMS 1 sheet boundaries, were plotted on both the NZMS 1 maps (latitude/longitude) and the character maps (row/column). The NZMS 1 maps were then enlarged to fit the character maps. In the field, the points were also used to align the overlay on the character map.

Figure 12.3 is a 1:63360 character map produced by photographically combining the lineprinter output with an overlay of roads, rivers and other topographical features.

12.4.20 Field visit 3: check of classification accuracy

The assessment of classification accuracy was conducted following the procedure outlined in Chapter 11.

The King Country was first divided into seven regions (see plate 1):

Kawhia	Whareorino	Waitaanga	Mangakino
Pureora	Tihoi	Tongariro.	

This ensured that any differences caused by climate, soils or terrain could be detected. Aerial photography, existing classification data and field staff from the district were used for ground checking.

Because of the time lapse since the Darfield–Eyrewell image was recorded, existing forest record maps, windthrow information (maps and moisture content records) and aerial photographs were used to check the accuracy of that classification map.

NB.The King Country indigenous forests are not rapidly changing and there was little change between the Landsat image (February 1976) used in plate 1 and the time of the field trips (throughout 1980). In Darfield–Eyrewell, where the forests are being intensively managed, substantial changes had occurred between image acquisition and field work.

12.5 Results

Results are reported under the following sections.
12.5.1   Classification accuracy
12.5.2   Ratioed and Principal Component channels for classification
12.5.3   Registration accuracy
12.5.4   Cost comparison
12.5.5   Classification maps
12.5.6   Summary of classification areas
12.5.7   Mean and standard deviation reports for classes.

12.5.1 Classification accuracy

12.5.1.1 King Country

Tables 12.5–12.11 show the number of pixels checked, and the numbers correct and incorrect for each class, for each region. The probabilities of a pixel being correctly classified with 99.9% confidence limits have been calculated following the techniques of Chapter 11. Although Chapter 11 recommends a sample size of greater than 50, samples larger than 30 were used because of the difficulty of locating and checking classes on the character map (of which figure 12.3 is a small part) without biasing the results. All class statistics were carried forward into the total regional summary shown in tables 12.12(a) and (b). A comparison of the class probabilities between the regions (tables 12.5–12.11) indicated that the results have been influenced by both aspect (sun/shade) and terrain.

The rougher terrains of the coastal regions (Whareorino and Waitaanga) have reduced the confidence levels of the podocarp forests accuracy values. Shadows cast by the terrain were confused with podocarp forest. The introduction of a 'shade–podocarp' class and further refinement of the podocarp class could solve this confusion in any following study.

In checking the Kawhia and Whareorino regions, assistance was gained from several members of the local district office. Interpretations appear to have some influence on the class probabilities shown in table 12.12(a).

In the King Country region, 0.5% ($12\,954/2\,429\,000 \times 100$) of the pixels were field checked, with $10\,745$ correctly classified. Thus the probability of a pixel being classified into its correct class, 999 times out of 1000, was 81.9% $\pm$ 0.5% ($\pm$ 0.5% being the measuring or counting error).

Conclusions

(i)   The forest scrub boundaries have been well defined (see plate 3).

(ii)   Where the terrain ranges from flat to undulating the classification defined the major indigenous species correctly.

(iii)   In steeper terrain, there was some misclassification of species. These misclassifications tend to be on the southwestern faces, in this southern hemisphere study, and appeared to be caused by terrain shadows. This indicated that the classes need further refinement.

(iv)   The classification of the exotic forest classes was poor due to impure training fields. This forest was relatively young and, being the first crop since the land was cleared, it contained considerable scrub. The survival rate of the exotic trees was therefore expected to be lower than normal. As no other training fields were available, the classes were included in classification but for trial purposes only.

(v)   Misclassifications in dense Kanuka–Manuka indicated that further refinement was still necessary because of the training fields or classes being too broad.

(vi)   The diversity of forest mixtures and canopy covers also affected the classification.

Table 12.5 Data for Kawhia region in King Country using nomenclature from §11.6.

Class	No pixels checked N	No correct P	No incorrect Q	Binomial st. dev. $s = (NPQ/N^2)^{1/2}$	$3s$	99.9% CL $100(P-3s)/N$ (%)
Water	—	—	—	—	—	—
Bare ground/urban	—	—	—	—	—	—
Scrub						
Fern	45	44	1	0.99	2.97	91.2
Kanuka/manuka	72	61	11	3.05	9.15	72.0
Manoao/tussock	9	1	8			
Hardwood scrub 1	10	10	0			
Hardwood scrub 2	43	46	2	1.38	4.14	87.2
Alpine/low scrub	—					
Broadleaf/smaller hardwoods	44	32	12	2.95	8.85	52.6
Indigenous forest						
Tawa	10	7	3			
Tawa 2	38	35	3	1.66	4.98	79.0
Tawa 3	281	193	88	7.77	23.31	60.4
Tawa/hardwood	38	37	1	0.99	2.97	89.6
Podocarp	302	183	119	8.49	25.47	52.2
Podocarp/hardwood 1	80	68	12	3.19	9.57	73.0
Podocarp/hardwood 2	46	32	14	3.12	9.36	49.2
Podocarp/hardwood 3	2	1	1			
Beech	8	0	8			
Kamahi	10	10	0			
Hardwood	18	18	0			

Table 12.5 (contd)

Class		No pixels checked N	No correct P	No incorrect Q	Binomial st. dev. $s = (NPQ/N^2)^{1/2}$	$3s$	99.9% CL $100(P-3s)/N$ (%)
Exotic forest							
P radiata	50	56	0	56	0.00	0.00	00.0
	56	5	0	5			
	62	—					
	70	—					
P contorta	59	—					
	69	2	0	2			
Ps menziesii	50	—	—	—			
	60	—	—	—			
	69	7	0	7			

Table 12.6 Data for Wharerino region in King Country with binomial standard deviation and CL as in table 12.5.

Class	No pixels checked N	No correct P	No incorrect Q	Binomial st. dev. s	$3s$	99.9% CL (%)
Water	212	207	5	2.21	6.63	94.5
Bare ground/urban	4	4	0			
Scrub						
Fern	106	97	9	2.87	8.61	83.4
Kanuka/manuka	43	43	0		0	
Manoao/tussock	—	—	—			
Hardwood scrub 1	75	75	0		0	
Hardwood scrub 2	104	97	7	2.56	7.68	85.9
Alpine/low scrub	—	—	—			
Broadleaf/smaller hardwoods	80	74	6	2.36	7.08	83.7
Indigenous forest						
Tawa	28	24	4			
Tawa 2	91	80	11	3.11	9.33	77.7
Tawa 3	439	404	35	5.68	17.04	88.2
Tawa/hardwood	52	50	2	1.39	4.17	88.1
Podocarp	345	162	183	9.27	27.81	38.9
Podocarp/hardwood 1	160	101	59	6.10	18.30	51.7
Podocarp/hardwood 2	48	31	17	3.31	9.93	43.9
Podocarp/hardwood 3	7	7	—			
Beech	7	6	1			
Kamahi	84	80	4	1.95	5.85	88.3
Hardwood	80	77	3	1.70	5.10	89.8

Table 12.6 (contd)

Class	No pixels checked N	No correct P	No incorrect Q	Binomial st. dev. s	3s	99.9% CL (%)
Exotic forest						
P radiata						
50	72	2	70	1.39	4.17	—
56	17	0	17			
62	—	—	—			
70	—	—	—			
P contorta						
59	1	0	1			
69	4	0	4			
Ps menziesii	50	1	0	1		
60	3	0	3			
69	13	0	13			

Table 12.7 Data for Waitaanga region in King Country with binomial standard deviation and CL as in table 12.5.

Class	No pixels checked N	No correct P	No incorrect Q	Binomial st. dev. s	$3s$	99.9% CL (%)
Water	114	109	5	2.19	6.57	89.9
Bare ground/urban	5	5	—			
Scrub						
Fern	235	230	5	2.21	6.63	95.1
Kanuka/manuka	146	134	12	3.32	9.96	85.0
Manoao/tussock	2	1	1			
Hardwood scrub 1	77	72	5	2.16	6.48	85.1
Hardwood scrub 2	140	128	12	3.31	9.93	84.3
Alpine/low scrub	47	41	6	2.29	6.87	72.6
Broadleaf/smaller hardwoods	102	94	8	2.72	8.16	84.2
Indigenous forest						
Tawa	14	14	—			
Tawa 2	71	63	8	2.66	7.98	77.5
Tawa 3	518	438	80	8.22	24.66	79.8
Tawa/hardwood	69	62	7	2.51	7.53	78.9
Podocarp	553	357	196	11.25	33.75	58.5
Podocarp/hardwood 1	173	145	28	4.84	14.52	75.4
Podocarp/hardwood 2	120	85	35	4.98	14.94	58.4
Podocarp/hardwood 3	5	4	1			
Beech	12	8	4			
Kamahi	39	37	2	1.38	4.14	84.3
Hardwood	60	56	4	1.93	5.79	83.7

Table 12.7 (contd)

Class		No pixels checked N	No correct P	No incorrect Q	Binomial st. dev. s	$3s$	99.9% CL (%)
Exotic forest							
P radiata	50	19	0	19			
	56	13	0	13			
	62	1	0	1			
	70	—	—	—			
P contorta	59	2	0	2			
	69	—	—	—			
Ps menziesii	50	1	0	1			
	60	2	0	2			
	69	11	0	11			

Table 12.8 Data for Pureora region in King Country with binomial standard deviation and CL as in table 12.5.

Class	No pixels checked N	No correct P	No incorrect Q	Binomial st. dev. s	3s	99.9% CL (%)
Water	87	81	6	2.36	7.08	85.0
Bare ground/urban	241	239	2	1.41	4.23	97.4
Scrub						
Fern	130	130	—			
Kanuka/manuka	254	238	16	3.87	11.61	89.1
Manoao/tussock	22	17	5			
Hardwood scrub 1	2	2	—			
Hardwood scrub 2	33	31	2	1.37	4.11	81.5
Alpine/low scrub	61	57	4	1.93	5.79	84.0
Broadleaf/smaller hardwoods	90	87	3	1.70	5.10	91.0
Indigenous forest						
Tawa	104	101	3	1.71	5.13	92.2
Tawa 2	80	77	3	1.70	5.10	89.9
Tawa 3	198	175	23	4.51	13.53	81.6
Tawa/hardwood	147	141	6	2.40	7.20	91.0
Podocarp	368	356	12	3.41	10.23	94.0
Podocarp/hardwood 1	367	365	2	1.41	4.23	98.3
Podocarp/hardwood 2	70	56	14	3.35	10.05	65.6
Podocarp/hardwood 3	6	5	1			
Beech	9	—	9			
Kamahi	69	40	29	4.10	12.30	40.1
Hardwood	72	54	18	3.67	11.01	59.7

Table 12.8 (contd)

Class		No pixels checked N	No correct P	No incorrect Q	Binomial st. dev. s	3s	99.9% CL (%)
Exotic forest							
P radiata	50	18	—	18			
	56	52	—	52			
	62	39	33	6	2.25	6.75	67.3
	70	13	1	12			
P contorta	59	8	0	8			
	69	46	37	9	2.69	8.07	62.9
Ps menziesii	50	27	22	5			
	60	40	31	9	2.64	7.92	57.7
	69	46	24	22	3.39	10.17	30.1

Table 12.9 Data for Mangakino region in King Country with binomial standard deviation and CL as in table 12.5.

Class	No pixels checked N	No correct P	No incorrect Q	Binomial st. dev. s	3s	99.9% CL (%)
Water	157	155	2	1.41	4.23	96.0
Bare ground/urban	97	95	2	1.40	4.20	93.6
Scrub						
Fern	13	13	0			
Kanuka/manuka	3	3	0			
Manoao/tussock	—	—	—			
Hardwood scrub 1	2	2	0			
Hardwood scrub 2	1	1	0			
Alpine/low scrub	2	2	0			
Broadleaf/smaller hardwoods	2	2	0			
Indigenous forest						
Tawa	15	0	15			
Tawa 2	16	0	16			
Tawa 3	9	0	9			
Tawa/hardwood	8	0	8			
Podocarp	37	0	37			
Podocarp/hardwood 1	11	0	11			
Podocarp/hardwood 2	—	—	—			
Podocarp/hardwood 3	—	—	—			
Beech	2	0	2			
Kamahi	26	17	9			
Hardwood	2	0	2			

Table 12.9 (contd)

Class		No pixels checked N	No correct P	No incorrect Q	Binomial st. dev. s	$3s$	99.9% CL (%)
Exotic forest							
P radiata	50	72	57	15	3.45	10.35	64.8
	56	254	224	30	5.14	15.42	82.1
	62	9	7	2			
	70	20	20	0			
P contorta	59	31	0	31			
	69	3	0	3			
Ps menziesii	50	11	0	11			
	60	2	0	2			
	69	2	0	2			

Table 12.10 Data for Tihoi region in King Country with binomial standard deviation and CL as in table 12.5.

Class	No pixels checked N	No correct P	No incorrect Q	Binomial st. dev. s	3s	99.9% CL (%)
Water	75	75	0			
Bare ground/urban	86	82	4	1.95	5.85	88.6
Scrub						
Fern	49	41	8	2.59	7.77	67.8
Kanuka/manuka	116	116	0			
Manoao/tussock	4	4	0			
Hardwood scrub 1	58	50	8	2.63	7.89	72.6
Hardwood scrub 2	19	17	2			
Alpine/low scrub	13	13	0			
Broadleaf/smaller hardwoods	39	39	0			
Indigenous forest						
Tawa	35	32	3	1.66	4.98	77.2
Tawa 2	43	35	8	2.55	7.65	63.6
Tawa 3	62	38	24	3.84	11.52	42.7
Tawa/hardwood	51	45	6	2.30	6.90	74.7
Podocarp	316	313	3	1.72	5.16	97.4
Podocarp/hardwood 1	147	144	3	1.71	5.13	94.5
Podocarp/hardwood 2	39	35	4	1.89	5.67	75.2
Podocarp/hardwood 3	18	18	0			
Beech	8	4	4			
Kamahi	51	45	6	2.30	6.90	74.7
Hardwood	29	29	0			

Table 12.10 (contd)

Class	No pixels checked N	No correct P	No incorrect Q	Binomial st. dev. s	$3s$	99.9% CL (%)
Exotic forest						
P radiata						
50	4	0	4			
56	29	0	29			
62	4	0	4			
70	—	—	—			
P contorta						
59	7	0	7			
69	3	0	3			
Ps menziesii	50	1	0	1		
60	—	—	—			
69	2	0	2			

Table 12.11 Data for Tongariro region in King Country with binomial standard deviation and CL as in table 12.5.

Class	No pixels checked N	No correct P	No incorrect Q	Binomial st. dev. s	3s	99.9% CL (%)
Water	1094	986	108	9.87	29.61	87.4
Bare ground/urban	84	75	9	2.83	8.49	79.2
Scrub						
Fern	98	87	11	3.12	9.36	79.2
Kanuka/manuka	90	52	38	4.69	14.07	42.1
Manoao/tussock	139	136	3	1.71	5.13	94.2
Hardwood scrub 1	27	27	0			
Hardwood scrub 2	41	41	0			
Alpine/low scrub	29	29	0			
Broadleaf/smaller hardwoods	103	98	5	2.18	6.54	88.8
Indigenous forest						
Tawa	3	2	1			
Tawa 2	28	27	1			
Tawa 3	77	58	19	3.78	11.34	60.6
Tawa/hardwood	83	66	17	3.68	11.04	66.2
Podocarp	172	153	19	4.11	12.33	81.8
Podocarp/hardwood 1	92	72	20	3.96	11.88	65.4
Podocarp/hardwood 2	109	80	29	4.61	13.83	60.7
Podocarp/hardwood 3	11	11	0			
Beech	23	23	0			
Kamahi	26	26	0			
Hardwood	36	36	0			

Table 12.11 (contd)

Class	No pixels checked N	No correct P	No incorrect Q	Binomial st. dev. s	$3s$	99.9% CL (%)
Exotic forest						
P radiata 50	8	0	8			
56	7	5	2			
62	—	—	—			
70	—	—	—			
P contorta 59	2	0	2			
69	—	—	—			
Ps menziesii 50	—	—	—			
60	—	—	—			
69	—	—	—			

Table 12.12(*a*) King Country: summary (per region). See § 11.6 for determination of CL for a classification.

Region	No pixels checked N	No correct P	No incorrect Q	99.9% lower confidence $100[(m-3e_m) - 3(s+3e_s)]/N$ (%)
Kawhia	1131	778	353	64.27
Whareorino	2076	1621	455	75.17
Waitaanga	2551	2083	468	79.21
Pureora	2699	2400	299	87.00
Mangakino	807	598	209	68.97
Tihoi	1308	1175	133	87.11
Tongariro	2382	2090	292	85.60
King Country Total	12 954	10 745	2209	81.93

Table 12.12(*b*) King Country: summary (per class). See § 11.6 for determination of CL for a classification.

Class	No pixels checked N	No correct P	No incorrect Q	99.9% lower confidence $100[(m-3e_m) - 3(s+3e_s)]/N$ (%)
Water	1739	1613	126	90.75
Bare ground/urban	517	500	17	94.04
Scrub				
Fern	676	642	34	92.15
Kanuka/manuka	724	647	77	85.53
Manoao/tussock	176	159	17	82.09
Hardwood scrub 1	251	238	13	89.80
Hardwood scrub 2	386	361	25	89.17
Alpine/low scrub	152	142	10	85.86
Broadleaf/smaller hardwoods	460	426	34	88.42
Indigenous forest				
Tawa	209	180	29	77.40
Tawa 2	367	317	50	80.13
Tawa 3	1584	1306	278	79.36
Tawa/hardwood	448	401	47	84.53
Podocarp	2093	1524	569	69.70
Podocarp/hardwood 1	1030	895	135	83.43
Podocarp/hardwood 2	432	319	113	66.55
Podocarp/hardwood 3	49	46	3	79.02
Beech	69	41	28	35.02
Kamahi	305	255	50	76.11
Hardwood	297	270	27	85.00

Table 12.12(*b*) (contd)

Class		No pixels checked *N*	No correct *P*	No incorrect *Q*	99.9% lower confidence $100[(m-3e_m) - 3(s+3e_s)]/N$ (%)
Exotic forest					
P radiata	50	249	59	190	14.01
	56	377	229	148	51.98
	62	53	40	13	50.14
	70	33	21	12	24.86
P contorta	59	51	0	51	0.00
	69	58	37	21	37.10
Ps menziesii	50	41	22	19	18.19
	60	47	31	16	35.78
	69	81	24	57	9.13

12.5.1.2 Darfield–Eyrewell

Whilst the same statistical approach was used in checking the Darfield–Eyrewell classification as in the King Country, the aims were different. Accuracy changes were observed by grouping classes and using three sets of channels (four MSS, eight MSS, fifteen optimum) in the classification (see §§12.4.15 and 12.4.17.)

Classes were grouped by selecting *P radiata* 1968, 1970 and 1972 age classes and adding ± 1 year. Hence classes 1967–9, 1969–71 and 1971–3 were defined. It was considered that if a stand of trees could be identified to within ± 1 year of their planting, the data would be acceptable. In obtaining the accuracy statistics presented in tables 12.16–12.19, no pixel was observed more than once.

Results, shown in tables 12.13–12.15, give the accuracy of each class using four MSS channels (table 12.13) of scene 2282–21 254, fifteen channels (MSS and synthetic) recommended via Divergence in scene 2282–21 254 (table 12.14) and a multitemporal approach using the eight MSS channels from scenes 2282–21 254 and 2192–21 265 (table 12.15).

The low accuracies achieved in classifying the forest classes, tables 12.13–12.15, suggested that the level of classification could not be achieved by this method. It was decided to group the forest classes into three-year age classes. Tables 12.16–12.18 show considerably improved results over those in tables 12.13–12.15 as a consequence.

Table 12.13 Darfield–Eyrewell classification using 4 MSS channels in scene reference number 2282–21 254.

Class	No pixels checked N	No correct P	No incorrect Q	Binomial st. dev. $s=(NPQ/N^2)^{1/2}$	99.9% CL $100(P-3s)/N$ (%)
Water	175	171	4	1.98	94.3
River gravel	61	60	1	0.99	93.5
P radiata 67	145	35	110	5.15	13.5
68	122	60	62	5.52	35.6
68cc†	122	10	112	3.03	0.7
69	178	115	63	6.38	53.8
69cc†	309	74	235	7.50	16.7
70	83	14	69	3.41	4.5
70cc†	228	163	65	6.82	62.5
71	241	129	112	7.74	43.9
72	333	229	104	8.46	61.2
73	252	126	126	7.94	40.5
P nigra 58	68	60	8	2.66	76.5
Mixed exotics	78	62	16	3.57	65.8
Serious windthrow	360	355	5	2.22	96.8
Moderate windthrow	113	73	40	5.08	51.1

†See table 12.1(*b*) for definitions.

Table 12.14 Darfield–Eyrewell classification using 15 channels in scene reference number 2282–21 254 with binomial standard deviation and CL as in table 12.13.

Class	No pixels checked N	No correct P	No incorrect Q	Binomial st. dev. s	99.9% CL (%)
Water	144	140	4	1.97	93.1
River gravel	64	62	2	1.39	90.4
P radiata 67	50	32	18	3.39	43.6
68	54	34	20	3.55	43.3
68cc	15	2	13	—	—
69	78	45	33	4.36	40.9
69cc	129	48	81	5.49	24.4
70	46	12	34	2.98	6.7
70cc	66	56	10	2.91	71.6
71	96	57	39	4.81	44.3
72	106	87	19	3.95	70.9
73	107	20	87	4.03	7.4
P nigra 58	57	48	9	2.75	69.7
Mixed exotics	67	46	21	3.80	51.6
Serious windthrow	204	197	7	2.60	92.7
Moderate windthrow	86	4	82	—	—

Table 12.15 Darfield–Eyrewell classification using 8 channels in scene reference number 2282–21 254, 2192–21 265 with binomial standard deviation and CL as in table 12.13.

Class	No pixels checked N	No correct P	No incorrect Q	Binomial st. dev. s	99.9% CL (%)
Water	92	90	2	1.40	93.3
River gravel	55	54	1	0.99	92.8
P radiata 67	61	40	21	3.71	47.3
68	92	56	36	4.68	45.6
68cc	45	3	42	—	—
69	102	66	36	4.83	50.5
69cc	166	77	89	6.43	34.8
70	77	24	53	4.06	15.3
70cc	129	90	39	5.22	57.6
71	366	281	85	8.08	70.2
72	271	243	28	5.01	84.1
73	195	116	79	6.86	48.9
P nigra 58	58	48	10	2.88	67.9
Mixed exotics	56	46	10	2.87	66.8
Serious windthrow	324	320	4	1.99	98.9
Moderate windthrow	—	—	—	—	—

Table 12.16 Darfield–Eyrewell classification using 4 mss channels in scene reference number 2282–21 254 (combined classes).

Class	No pixels checked N	No correct P	No incorrect Q	Binomial st. dev. s	99.9% CL (%)
P radiata 1967–9	332	215	117	8.70	56.9
P radiata 1969–71	265	235	30	5.16	82.8
P radiata 1971–3	405	303	102	8.74	68.3
Other species	138	122	16	3.76	80.2
Windthrow	284	280	4	1.99	96.5
Water	175	171	4	1.98	94.3
River gravel	61	60	1	0.99	93.5

Table 12.17 Darfield–Eyrewell classification using 15 channels in scene reference number 2282–21 254 (combined classes).

Class	No pixels checked N	No correct P	No incorrect Q	Binomial st. dev. s	99.9% CL (%)
P radiata 1967–9	210	163	47	6.04	69.0
P radiata 1969–71	251	229	22	4.48	85.9
P radiata 1971–3	165	140	25	4.61	76.5
Other species	71	60	11	3.05	71.6
Windthrow	382	324	58	7.01	79.3
Water	141	140	1	1.00	97.2
River gravel	34	32	2	1.37	82.0

Table 12.18 Darfield–Eyrewell classification using 8 MSS channels in scene reference number 2282–21 254, 2192–21 265 (combined classes).

Class	No pixels checked N	No correct P	No incorrect Q	Binomial st. dev. s	99.9% CL (%)
P radiata 1967–9	337	276	61	7.07	75.6
P radiata 1969–71	177	165	12	3.34	87.5
P radiata 1971–3	283	240	43	6.04	78.4
Other species	84	64	20	3.90	62.3
Windthrow	324	320	4	1.99	96.9
Water	92	90	2	1.40	93.3
River gravel	55	54	1	0.99	92.8

Table 12.19 Overall accuracy changes for Darfield–Eyrewell. See § 11.6 for determination of CL for a classification.

Bands used for classification	No pixels checked N	No correct P	No incorrect Q	99.9% lower confidence $100[(m-3e_m) - 3(s+3e_s)]/N$ (%)
4 MSS Scene 2282–21 254	2868	1736	1132	57.6
15 bands Scene 2282–21 254	1369	890	479	60.8
8 MSS Scene 2282–21 254 Scene 2192–21 265	2089	1554	535	71.3
Grouped classes 4 MSS Scene 2282–21 254	1660	1386	274	80.5
15 bands Scene 2282–21 254	1254	1088	166	83.6
8 MSS Scene 2282–21 254 Scene 2192–21 265	1352	1209	143	86.7

Table 12.19 displays the overall accuracy changes. A summary of this is shown below giving the probability of a pixel being classified into its correct class, 999 times out of 1000:

(i)   4 MSS   57.6%   ±0.5%
(ii)   15 channels   60.8%   ±0.5%
(iii) 8 MSS   71.3%   ±0.5%
(iv)   4 MSS, grouped classes   80.5%   ±0.5%
(v)   15 channels, grouped classes   83.6%   ±0.5%
(vi)   8 MSS, grouped classes   86.7%   ±0.5%.

Possibly even better results could have been achieved using synthetic channels, created from the eight MSS. Careful planning would have been required, however, because of the large number of possible bands.

Conclusions

(i)   Serious wind damage was well defined (see plate 2). It was shown clearly in each of the four MSS (96.8%), eight MSS (98.9%) and fifteen-channel (92.7%) classifications (refer to tables 12.13–12.18).
(ii) A satisfactory classification was achieved by grouping yearly age classes into three-year age classes.

(iii)   The best forestry classifications were achieved by using either the synthetic and some of the MSS bands of one scene which had been selected via the channel separability approach or the eight MSS bands from two scenes.

(iv)   The 15 best channels classification was an improvement on the four MSS band classification. Use of multitemporal data (two scenes) made a further improvement in classification accuracy.

12.5.2   *Use of Ratioed and Principal Component channels in classification*

Considerable discussion and thought was given to the merits of Ratioed and Principal Component bands. The main concern being 'does a derived/synthetic band increase separability between classes'. Divergence showed that Ratioed and Principal Component channels aided the separability necessary for the desired level of classification.

Method and results of King Country Divergence

Each time further channels were added to the channel set a classification was carried out over the training fields with a 0% threshold. This indicated the improvement in separation gained by adding the channel(s). The improvement could be readily seen by reference to the percentage correct value in the classification summary report.

(i)   The first classification was made using the four Landsat MSS bands. This formed a baseline. A mean of 78.7% (standard deviation 19.8%) of the pixels, within the training areas over all classes, were classified correctly.

(ii)   Average Divergence (D_{ave}) listed the best four bands from the twenty available channels (table 12.2) and ranked the separation between class pairs. Separation was indicated by a value ranging from 999 (high separability) to 1 (low separability). The recommended four channels were MSS 6, '5/4', '6/4' and '6/5' for the forestry application. A list was compiled of the class pairs with Divergence values less than 900. The lowest divergence value found was 429 for the class pair tawa 3–hardwood. Interclass pair Divergences (D_{ij}) were then used for each of these class pairs to recommend the best channels for separability (see table 12.3, column (*a*) previously). A further seven channels were selected making a total of eleven channels: 6, '5/4', '6/4', '6/5' plus: 4, '4/5', '4/6', 5, '5/7', '7/4', '7/5'.

(iii)   A classification of training fields (using the eleven channels) gave a mean of 80.5% (standard deviation 18.2%) for correctly classified pixels.

(iv)   The summaries of the classification reports were studied to locate unacceptable mixtures. Interclass pair Divergence (D_{ij}) was used to select further channels to remove the unacceptable mixtures, as follows.

Unacceptable mixtures between:	New D_{ij}	Recommended channels
Tawa 2–*Ps menziesii* 60	999	PC1, PC2, '5/7', '7/5'
Tawa 2–*Ps menziesii* 69	999	PC1, '6/5', 5, 6
Podocarp hardwood 1–beech	999	PC1, 4, '4/5', '5/7'
Alpine/low scrub–*Ps menziesii* 60	999	PC1, 4, '4/6', '5/6'
Tawa–beech	999	PC1, 4, '4/5', '6/4'
Tawa 2–beech	999	PC1, '4/7', 4, 7

Note: (i) PC stands for Principal Component. (ii) While tawa 3–hardwood had the lowest Divergence value, it was considered to be an acceptable mixture (in table 12.3).
Channels PC1, '5/6', 7 were included with the eleven channels making a total of fourteen channels.

(v)   Classification of training fields was then carried out using the fourteen channels for the King Country study: PC1, 4, '4/5', '4/6', 5, '5/4', '5/6', '5/7', 6, '6/4', '6/5', 7, '7/4', '7/5'. A mean accuracy of 85.4% with a standard deviation of 19.5% was achieved for correctly classified pixels. This was considered acceptable, although the 'beech' and 'podocarp–hardwood 3' classes had rather low accuracy values. Nevertheless, because of time constraints it was decided to proceed and classify the whole King Country area using the fourteen channels.

Conclusion
The manner in which Divergence selected the channels has indicated the need for Ratioed and Principal Component channels.

12.5.3 Registration accuracy
Registration accuracy is influenced by a combination of factors:

(i)   identification of ground control points (GCP);

(ii)   reading errors of GCP latitude and longitude values;
(iii)   locating GCP on the interactive monitor of the computer analysis system;
(iv)   registration to the preferred projection;
(v)   the difference between the computer system's projection and that of the map base used for checking;
(vi)   the identification of the assessment points on both the computer generated product and the checking maps;
(vii)   reading errors of map coordinate values for assessment points.

The accuracy of the registered King Country image was found to be ± 125 m. This was estimated by measuring 35 assessment points on MSS band 4 of the output product. This represents approx ± 1 mm at 1:100 000 scale and ± 2 mm at 1:50 000, or 1.5 Landsat 80 m resampled pixels.

Ten points measured on the Darfield–Eyrewell data gave an accuracy of ± 61 m. Work on a limited area with 50 further points has indicated that the accuracy could be ± 1 (40 m resampled) pixel.

Chapter 8 provides a more complete report on registration.

Conclusion

The registration accuracy (± 125 m for King Country and ± 61 m for Darfield–Eyrewell) is acceptable for forestry purposes.

The better accuracy of the Darfield–Eyrewell image is attributed firstly to 40 m resampling of the image carried out during Registration, as against the 80 m for King Country and secondly, the use of more well-defined GCP (mainly road intersections), which was possible in the managed exotic forest environment.

12.5.4 Cost comparison

In 1977, part of the King Country region was surveyed by conventional methods for a land use report (*King Country Land Use Study* 1977). A comparison was made between the forestry part of this survey and this computer-based study (tables 12.20(*a*) and (*b*)). (Products from the *two* studies may be compared by reviewing plate 3 and figure 12.4.) Only the costs (1980 values) of gathering the information were considered and not those of final map/report preparation.

The cost of the conventional survey was NZ$20/km². The Landsat–computer system survey cost NZ$3.5/km². Furthermore, a man-year was required to survey 970 km² by conventional survey. The Landsat–computer method surveyed 14 200 km² (of land) in a man-year. Thus, Landsat–computer gives a 93% saving in manpower and an 82% reduction in total costs over conventional survey for this study.

12.5.5 Classification maps

12.5.5.1 Character maps

The lineprinter character map output and a road/river overlay (NZMS 1) were both photographically reproduced at 1:50 000 and combined. Topographic information in the form of hill shading was also included to assist in the location of shadow/sunlit areas (§ 12.5.1.1, conclusion (iii)). A transparent film copy was made to allow relatively inexpensive copies to be made, as in figure 12.3.

12.5.5.2 Colour maps

The colour maps produced on the system and written out on the New Zealand Optronics C4300 Colorwrite instrument and photographically combined with a road/river overlay were as follows.

King Country (14-channel classification showing the following classes).(i) Scrub, (ii) *P radiata*, (iii) *Ps menziesii*/*P contorta*, (iv) podocarp forest (plates 1 and 3), (v) tawa, (vi) other indigenous forest, (vii) general class map.

Darfield–Eyrewell (15-channel classification showing the following classes). (i) Windthrown forest (plate 2), (ii) *P radiata*, 1960s planting, (iii) *P radiata*, 1970s planting, (iv) other forest classes.

The scale of these can obviously be varied to suit user needs. Plates 1 and 3 are examples of this product at scales 1:1 000 000 and 1:63 360 respectively.

12.5.5.3 Comparison of classification products

A comparison of conventional and Landsat–ERMAN classification products can be seen in plate 3 and figure 12.4. Special attention is drawn to the finer detail of the Landsat–ERMAN product as against the 'broad-brush' approach of the conventional map.

12.5.6 Summary of classification areas

Table 12.21 gives the total areas (in hectares) for each class within the King Country region. These were obtained from the class summary report for the total classified area. This report presented the total number of pixels (80×80 m^2) classified into each class. These totals were then converted to hectares (1 pixel = 0.64 hectares).

Table 12.20(*a*) Summary of costs (at 1980 NZ$ levels) for King Country, conventional survey (forest information only). The area surveyed was 4375 km². Aerial photography has not been included in these costs because suitable photography was already available for this survey. However, it is estimated that aerial photography would have cost NZ$26 800.

Expense	Cost (NZ$)
Staff—$1\frac{1}{2}$ man-years	20 000
Student staff—3 man-years	24 000
Travel, accommodation, vehicles	15 000
Stores	2000
Total cost	61 000

Table 12.20(*b*) Summary of costs (at 1980 NZ$ levels) for King Country, Landsat–ERMAN survey. The area surveyed was 16 400 km² (including 1500 km² of water).

Expense	Cost (NZ$)
Staff, remote sensing officer—1 man-year	17 000
other—18 man-days	700
Travel within New Zealand	1000
Vehicles/aircraft for ground truth	2100
Stores (maps, airphotos, Landsat prints, etc	1900
Landsat data (1 tape)	200
Computer expenses—ancillary system support	12 100
ERMAN, 4.3 CPU hours at New Zealand rates	19 000
Travel to Australia	3000
Total cost	57 000

Comments (i) The ERMAN costs are dependent upon total computing time (CPU time) required. This in turn depends markedly on the amount of use of the Divergence option and on the number of channels and classes used in the classification. (ii) The Landsat data were purchased as a Computer Compatible Tape (CCT) for US$200.

12.5.7 Mean and standard deviation reports for classes

Table 12.22 shows the means and standard deviations, for each class, in all available channels. This information was computed after the training fields had been finalised and their statistics computed.

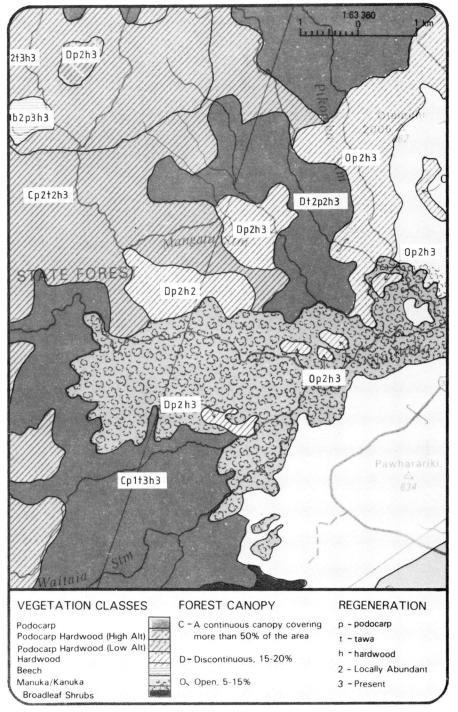

Figure 12.4 Black and white reproduction of forest and scrub type map; scale = 1:63 360; sheet 2 (NZMS 288, King Country land use study).

12.6 Conclusions

Increasing world demands for forest products together with local environmental pressures have led to more careful planning and monitoring of our forest resources. Satellite and computer processed imagery can make a valid and impartial contribution to the management process, especially given the rising costs of monitoring forest resources and declining staff resources.

A monitoring procedure has been developed and tested in two areas of New Zealand. The two trial areas represented different types of vegetation, terrain and degree of monitoring difficulty. The King Country ranges from flat pumice land to steeply dissected mudstone with mainly indigenous forest, scrubs and rangelands. It has proven difficult to survey by conventional means because of its mixed vegetation, rough terrain and lack of roading. Darfield–Eyrewell is a flat area on the Canterbury alluvial plains which supports highly productive cropping, with exotic forest plantations occurring on poorer soils or as wind shelter belts.

This method of forest type survey has achieved acceptable map registration ($\pm$ 125 m King Country, $\pm$ 61 m Darfield–Eyrewell) and classification accuracies (80–90%).

The cost of the conventional King Country survey was NZ$20/km^2, while the Landsat–ERMAN survey cost NZ$3.5/km^2. Furthermore, one man-year was required to survey 970 km^2 by conventional survey. The Landsat–ERMAN method surveyed 14 200 km^2 in a man-year. Thus, Landsat–ERMAN gives a 93% saving in manpower and an 82% reduction in total costs over conventional survey. The technique therefore appears to have potential for forest management.

The factors limiting a more accurate classification were the variety of vegetation types and the use of impure training fields. Further refinement of some classes would improve the accuracy of classification.

This type of survey relies totally on good ground truth data. In achieving this, approximately 90% of the effort was involved in gathering data with the remaining time computing.

This study suggests that the Landsat–computer system form of classification will be greatly improved by higher sun angle imagery and increased spectral and spatial resolution data. However, the greater the spatial resolution, the greater the number of micro-environmental variables to be considered, such as tree shadows and tree conditions. These extra variables could cause problems in this level of classification but must necessarily be considered if finer details are to be mapped.

The major exotic forests within New Zealand are intensively managed, well documented and have good road access. The indigenous forests are, however, documented to varying degrees and generally road access is poor or non-existent. Satellite information will therefore be of great value in:

(i) forming a base indigenous forest survey throughout the country. It will also provide for the continual monitoring of changes in vegetation boundaries, erosion, fire, wind damage and disease status;

(ii) locating private exotic woodlots, for example, shelter belts and farm forests, to form and maintain a national inventory;

(iii) monitoring of the exotic forest operations once satellite data become more readily available and spatial/spectral resolution improves.

Table 12.21 Summary of total areas in King Country Region.

Class	Hectares	Class	Hectares
Scrub		*Indigenous forest*	
Fern	124 200	Tawa	8100
Kanuka/manuka	52 000	Tawa 2	22 000
Monoao/tussock	9900	Tawa 3	103 900
Hardwood scrub 1	49 700	Tawa/hardwood	26 900
Hardwood scrub 2	31 800	Podocarp	92 000
Alpine/low scrub	33 100	Podocarp/hardwood 1	48 000
Broadleaf/small hardwoods	31 000	Podocarp/hardwood 2	28 000
Exotic forest		Podocarp/hardwood 3	3400
P radiata 50	6600	Beech	3200
P radiata 56	7900	Kamahi	15 000
P radiata 62	1100	Hardwoods	18 200
P radiata 70	600	*Other*	
P contorta 59	1900	Water	126 800
P contorta 69	1700	Bare ground/urban	26 300
Ps menziesii 50	1200		
Ps menziesii 60	1400		
Ps menziesii 69	5900		
		Threshold total	683 300
		Total area	1 565 100

Comments
(i) Threshold values placed on each class were those listed in table 12.4 and used in the classification map production.
(ii) Table should be read in conjunction with the class confidence of table 12.12(*b*).
(iii) Calculations to nearest 100 hectares.
(iv) The classes have been grouped as in table 12.12(*b*).

Table 12.22 Mean and standard deviation report for classes in all available channels for the King Country study. The figures in brackets give the number of pixels in training areas.

Channel	Channel number	Fern (35)		Kanuka/manuka (57)		Monoao/tussock (68)		Hardwood scrub 1 (55)		Hardwood scrub 2 (27)	
		Mean	St. dev.	Mean	St. dev.	Mean	St. dev.	Mean	St. dev.	Mean	St. dev.
PC1	1	26.69	2.40	15.37	1.29	14.91	0.64	29.31	2.13	23.85	1.03
PC2	2	146.54	1.99	137.46	1.05	136.66	0.75	148.49	1.56	144.33	0.83
PC3	3	117.03	0.45	117.28	0.49	117.15	0.36	117.49	0.63	117.52	0.51
PC4	4	127.23	0.43	127.16	0.37	126.99	0.12	127.91	0.55	127.74	0.45
4	5	12.29	0.62	10.61	0.56	10.56	0.53	13.45	1.15	12.19	0.48
'4/5'	6	26.91	1.46	27.86	1.60	25.84	1.64	30.95	2.50	30.07	1.59
'4/6'	7	7.34	0.64	12.25	1.30	12.91	1.46	7.07	0.69	8.07	0.38
'4/7'	8	6.37	0.73	12.04	1.59	12.44	0.97	6.07	0.69	7.11	0.58
5	9	13.34	0.54	11.04	0.46	11.93	0.26	12.78	0.90	11.81	0.40
'5/4'	10	31.97	1.65	30.26	1.70	32.72	1.40	27.96	2.17	28.26	1.13
'5/6'	11	7.94	0.87	12.77	1.28	14.51	1.14	6.80	0.62	7.85	0.53
'5/7'	12	6.36	0.97	12.46	1.56	14.01	1.15	5.73	0.56	6.96	0.52
6	13	50.37	5.25	26.07	2.82	24.54	1.73	55.82	4.57	44.26	2.33
'6/4'	14	120.69	9.51	71.40	7.27	67.69	6.42	123.53	9.63	107.04	4.79
'6/5'	15	111.89	11.28	68.82	7.01	60.40	4.33	129.36	9.16	110.19	6.66
'6/7'	16	27.09	1.09	29.89	2.22	29.47	1.97	26.47	1.07	26.96	1.26
7	17	28.40	3.35	12.81	1.78	12.12	0.74	32.18	2.76	24.85	1.54
'7/4'	18	136.14	12.87	70.14	8.98	66.84	5.45	142.49	11.82	120.41	6.87
'7/5'	19	126.29	14.17	67.75	8.99	59.94	4.04	148.98	10.06	124.04	8.54
'7/6'	20	34.74	1.38	29.88	2.42	30.00	2.06	35.87	1.40	34.56	1.53

Table 12.22 (contd)

Channel	Channel number	Alpine/ low scrub (22)		Broadleaf/ smaller hardwoods (21)		P radiata 1950 (49)		P radiata 1956 (51)		P radiata 1962 (25)	
		Mean	St. dev.	Mean	St. dev.	Mean	St. dev.	Mean	St. dev.	Mean	St. dev.
PC1	1	18.86	0.94	19.43	0.68	6.86	1.40	13.47	1.51	16.12	0.88
PC2	2	141.09	0.81	142.10	0.77	133.73	1.13	138.96	1.22	140.68	0.69
PC3	3	117.45	0.51	117.57	0.51	118.47	0.50	118.31	0.51	118.88	0.53
PC4	4	127.73	0.46	127.57	0.51	127.86	0.41	127.86	0.35	127.44	0.51
4	5	10.95	0.21	10.43	0.51	7.65	0.66	8.45	0.58	8.56	0.71
'4/5'	6	30.82	1.65	30.86	2.92	41.51	6.61	37.78	4.23	34.12	4.74
'4/6'	7	9.36	0.58	8.43	0.51	20.84	4.42	10.20	1.66	8.68	0.85
'4/7'	8	8.45	0.60	7.76	0.70	26.31	10.91	9.37	2.48	6.80	0.82
5	9	10.36	0.58	9.81	0.75	4.94	0.69	6.18	0.74	7.08	0.86
'5/4'	10	27.23	1.77	27.38	2.58	18.04	3.28	20.45	2.58	23.36	3.21
'5/6'	11	8.77	0.69	8.00	0.71	13.29	3.10	7.20	1.13	7.16	0.85
'5/7'	12	8.00	0.44	7.33	0.80	16.76	7.02	6.65	1.61	5.44	0.71
6	13	34.64	2.01	36.24	1.67	11.00	2.78	24.82	3.17	29.28	1.81
'6/4'	14	92.50	5.16	101.14	6.01	40.47	9.73	83.73	10.55	98.24	9.18
'6/5'	15	97.27	6.00	107.43	9.74	59.27	14.83	110.88	13.96	116.72	12.24
'6/7'	16	27.82	1.59	28.24	1.18	35.63	10.71	27.37	2.56	24.44	1.29
7	17	18.64	1.53	19.33	1.24	4.33	2.17	13.35	2.07	17.88	1.01
'7/4'	18	99.41	7.75	108.14	9.17	31.53	15.42	90.08	13.87	119.92	10.81
'7/5'	19	104.45	6.81	114.52	10.32	46.20	22.84	119.08	17.94	142.76	14.79
'7/6'	20	32.91	1.80	32.81	1.25	21.67	6.77	32.65	2.95	37.48	1.78

Table 12.22 (contd)

Channel	Channel number	P radiata 1970 (36)		P contorta 1959 (19)		P contorta 1969 (20)		Ps menziesii 1950 (18)		Ps menziesii 1960 (37)	
		Mean	St. dev.	Mean	St. dev.	Mean	St. dev.	Mean	St. dev.	Mean	St. dev.
PC1	1	21.69	1.37	13.84	0.69	17.00	0.56	18.78	1.06	24.46	1.07
PC2	2	145.31	1.14	139.11	0.46	141.45	0.51	142.67	0.91	147.16	1.01
PC3	3	118.67	0.59	118.37	0.50	118.30	0.57	118.28	0.57	118.11	0.31
PC4	4	127.83	0.38	127.89	0.32	127.95	0.22	127.94	0.42	127.95	0.23
4	5	9.72	0.51	8.84	0.37	9.00	0.32	9.61	0.92	10.43	0.50
'4/5'	6	35.47	3.45	36.74	3.62	35.55	3.32	35.72	4.69	33.22	1.49
'4/6'	7	6.78	0.72	10.26	0.73	8.50	0.61	7.72	0.96	6.19	0.46
'4/7'	8	5.47	0.56	8.95	0.71	7.35	0.67	6.61	0.85	5.30	0.46
5	9	7.78	0.80	6.74	0.65	7.15	0.81	7.61	0.78	9.03	0.50
'5/4'	10	22.86	2.37	21.58	2.36	22.45	2.26	22.72	3.12	25.08	1.16
'5/6'	11	5.36	0.59	7.68	0.75	6.65	0.93	6.11	0.68	5.49	0.56
'5/7'	12	4.25	0.44	6.74	0.73	5.60	0.75	5.22	0.73	4.51	0.56
6	13	41.89	3.09	25.53	1.22	32.00	1.34	35.89	2.22	48.00	2.42
'6/4'	14	124.89	11.21	82.84	6.82	102.20	5.46	108.56	12.00	134.08	8.20
'6/5'	15	153.19	13.07	105.74	5.99	126.50	12.88	133.94	12.14	153.24	9.90
'6/7'	16	25.03	1.06	26.47	1.43	26.90	1.45	26.00	1.19	25.89	0.88
7	17	25.28	1.61	14.16	0.83	17.75	0.91	20.67	1.19	28.03	1.61
'7/4'	18	150.83	11.87	91.74	6.94	113.25	7.19	125.17	12.78	156.84	11.90
'7/5'	19	185.00	15.38	117.63	10.51	140.40	15.00	154.44	14.78	178.89	13.75
'7/6'	20	37.39	1.81	33.79	1.87	34.15	2.01	35.39	1.61	36.22	1.11

Table 12.22 (contd)

Channel	Channel number	Ps menziesii 1969 (16)		Tawa (18)		Tawa 2 (21)		Tawa 3 (32)		Tawa/ hardwood (23)	
		Mean	St. dev.	Mean	St. dev.	Mean	St. dev.	Mean	St. dev.	Mean	St. dev.
PC1	1	23.87	2.09	13.78	1.35	16.29	0.85	20.56	1.74	15.65	1.23
PC2	2	145.75	2.21	138.28	0.89	139.81	0.68	142.91	1.30	139.13	0.97
PC3	3	117.69	0.79	118.06	0.42	118.00	0.32	117.75	0.51	117.91	0.51
PC4	4	127.56	0.51	127.50	0.51	127.81	0.40	127.81	0.40	127.39	0.50
4	5	10.75	0.45	8.83	0.51	9.76	0.54	10.97	0.47	9.52	0.51
'4/5'	6	29.62	2.33	32.72	3.29	32.29	2.67	32.25	2.76	29.57	2.25
'4/6'	7	6.94	1.06	10.78	1.26	10.00	0.63	8.53	0.92	10.39	0.89
'4/7'	8	5.87	1.15	9.78	1.06	8.76	0.83	7.53	1.02	9.13	0.76
5	9	10.62	1.09	7.67	0.59	8.62	0.50	9.87	0.79	9.17	0.58
'5/4'	10	28.62	2.66	24.44	2.23	25.43	2.25	26.16	2.33	27.48	1.73
'5/6'	11	6.94	1.18	9.28	0.75	8.71	0.78	7.59	0.76	9.87	0.97
'5/7'	12	6.00	1.41	8.39	0.50	7.86	0.79	6.66	0.79	8.87	0.63
6	13	45.56	5.10	24.61	2.70	29.38	1.80	38.41	3.60	27.83	2.74
'6/4'	14	123.75	13.95	79.67	8.40	87.14	5.06	102.44	9.85	84.30	7.46
'6/5'	15	126.44	20.34	90.56	6.71	97.57	7.98	112.78	9.57	87.22	7.72
'6/7'	16	26.50	1.55	27.56	1.82	27.33	1.24	26.97	1.60	27.30	1.87
7	17	26.06	3.49	13.11	1.45	16.05	1.24	21.50	2.40	15.09	1.12
'7/4'	18	141.87	19.99	85.00	9.08	95.14	6.99	114.66	12.37	91.43	6.51
'7/5'	19	145.00	27.88	96.39	7.81	106.52	9.27	126.25	11.89	94.70	7.55
'7/6'	20	35.25	2.29	32.50	1.98	33.67	1.39	34.47	1.95	33.35	1.87

Table 12.22 (contd)

Channel	Channel number	Podocarp (144)		Podocarp/hardwood 1 (22)		Podocarp/hardwood 2 (28)		Podocarp/hardwood 3 (17)		Beech (45)	
		Mean	St. dev.	Mean	St. dev.	Mean	St. dev.	Mean	St. dev.	Mean	St. dev.
PC1	1	11.11	0.91	14.00	0.87	13.46	1.10	17.35	0.79	14.18	0.58
PC2	2	135.85	0.74	138.05	0.72	136.89	1.13	140.65	0.79	138.18	0.58
PC3	3	118.50	0.52	118.00	0.69	118.00	0.54	117.71	0.47	118.02	0.40
PC4	4	127.56	0.50	127.64	0.49	127.54	0.51	127.71	0.47	127.96	0.21
4	5	8.85	0.37	9.55	0.51	9.79	0.42	9.71	0.47	9.89	0.32
'4/5'	6	34.56	3.45	33.45	1.92	30.75	2.08	30.82	1.88	34.67	0.95
'4/6'	7	14.62	1.57	11.59	0.73	13.04	1.73	9.12	0.86	11.67	0.74
'4/7'	8	13.33	1.87	10.68	0.89	12.36	1.85	7.94	0.24	10.64	0.74
5	9	7.25	0.69	8.05	0.21	9.14	0.45	9.00	0.00	8.00	0.00
'5/4'	10	23.09	2.31	24.05	1.33	26.89	1.71	26.59	0.94	23.22	0.64
'5/6'	11	11.83	1.54	9.73	0.98	12.36	1.64	8.47	0.72	9.49	0.69
'5/7'	12	10.64	1.76	8.82	0.66	11.71	1.78	7.47	0.51	8.67	0.48
6	13	18.10	1.95	24.59	2.15	22.54	2.66	32.00	2.09	24.98	1.57
'6/4'	14	58.42	6.33	74.18	5.11	66.50	8.75	95.47	6.64	72.84	5.54
'6/5'	15	70.37	8.36	86.64	7.87	70.82	8.89	102.18	6.61	88.40	5.67
'6/7'	16	27.04	2.61	27.77	2.33	28.39	2.15	27.35	1.77	27.49	1.25
7	17	9.65	1.35	13.00	1.07	11.57	1.69	17.47	0.62	13.24	0.61
'7/4'	18	62.31	8.64	78.50	7.14	68.36	10.66	103.82	4.45	77.36	4.87
'7/5'	19	75.07	11.48	91.59	8.10	72.71	11.14	111.35	4.23	93.71	4.26
'7/6'	20	31.90	3.11	32.27	2.96	31.07	2.62	33.82	2.07	32.31	1.58

Table 12.22 (contd)

Channel	Channel number	Kamahi (39)		Hardwood (44)		Water (9117)		Bare ground/ urban (141)	
		Mean	St. dev.	Mean	St. dev.	Mean	St. dev.	Mean	St. dev.
PC1	1	20.87	1.58	23.25	1.73	3.33	0.51	23.86	2.77
PC2	2	144.05	1.28	144.64	1.43	129.87	0.36	134.32	2.55
PC3	3	117.92	0.48	117.64	0.57	119.33	0.47	114.37	1.06
PC4	4	127.87	0.34	127.36	0.49	128.06	0.34	126.72	0.78
4	5	9.87	0.52	11.00	0.43	8.78	0.66	20.33	3.53
'4/5'	6	31.64	2.21	29.11	2.55	52.14	7.84	24.50	2.30
'4/6'	7	7.31	0.83	7.59	0.76	167.99	72.58	18.23	3.65
'4/7'	8	6.23	0.48	6.41	0.62	140.55	10.63	21.38	6.60
5	9	8.95	0.39	11.05	0.81	4.44	0.69	25.13	4.13
'5/4'	10	26.10	1.50	29.20	2.74	14.13	2.60	37.42	3.36
'5/6'	11	6.54	0.60	7.45	0.73	88.91	44.11	22.62	4.28
'5/7'	12	5.67	0.62	6.41	0.73	71.10	11.09	26.50	7.61
6	13	40.03	3.38	43.52	3.78	1.01	0.93	34.35	4.86
'6/4'	14	117.46	9.85	115.73	9.54	3.03	2.87	52.19	10.06
'6/5'	15	128.31	9.91	115.52	9.89	5.58	5.27	42.50	8.30
'6/7'	16	26.64	1.01	26.57	1.11	16.10	14.92	35.37	4.28
7	17	22.64	2.05	24.82	2.21	0.00	0.00	14.64	3.12
'7/4'	18	132.97	10.55	131.98	10.88	0.00	0.00	44.95	13.06
'7/5'	19	145.21	12.47	131.84	12.52	0.00	0.00	36.44	10.47
'7/6'	20	34.92	1.56	35.18	1.33	0.00	0.00	25.94	3.37

12.7 References

Aerial Photographs Survey Nos 2974, 2920, 3838 : 1976; 5014 : 1977; 5147, 5410 : 1979 (Hastings, NZ: New Zealand Aerial Mapping)

Beach D 1980 Informal communication IBM, Sydney

Dale R W 1981 Informal communication NZ Forest Service, Auckland

DSIR 1973 *Geological Survey and Soil Maps* (Wellington, NZ: NZ Geological Survey and Soil Bureau)

FSMS 6 1979 *Ecological Forest Class Maps* (Wellington, NZ: NZ Forest Service Mapping)

IBM 1976 *Earth Resources Management II (ER-MAN II) User's Guide* Program No 5790-ARB Doc. No SB 11-5008-0 (Brussels: IBM)

Joyce A T 1978 *Procedures for Gathering Ground Truth Information for a Supervised Approach to a Computer-Implemented Land Cover Classification of Landsat-acquired Multispectral Scanner Data* NASA Ref. Publ. 1015 (Washington, DC: NASA)

King Country Land Use Study 1977 (Wellington, NZ: Govt Printer)

NZ Forest Service 1955 *The National Forest Survey of New Zealand 1955* ed. S E Masters, J T Holloway and P J McKelvey (Wellington, NZ: Govt Printer)

NZFS/FP Ltd† *Exotic Plantation Record Maps* (Wellington, NZ: NZ Forest Service and Auckland, NZ: Forest Products Ltd)

NZMS 1 various dates 1971–82 (Wellington, NZ: Department of Lands and Survey)

NZMS 288 1977 (Wellington, NZ: Department of Lands and Survey)

Nicholls J L 1976 A revised classification of the North Island indigenous forests *NZ J. Forestry* **21** 105

Swain P H and Davis S M 1978 (ed.) *Remote Sensing: The Quantitative Approach* (New York: McGraw-Hill)

†Record maps are continuously updated.

13 | Land Cover Mapping from Landsat

13.1 Introduction

IBM's ERMAN II package was utilised to produce land cover classification maps from existing Landsat II imagery for use in the New Zealand Land Inventory Mapping Programme. The area chosen for study was New Zealand's central North Island King Country region, an area which had been difficult to assess and map using conventional methods. This chapter sets out the steps taken during the project: the collection of ground truth, preparation of enhanced Landsat imagery, registration, classification of these data into fifteen land cover classes, and the preparation of final classification maps. Detailed registration results and classification accuracies plus cost comparisons between ground and computer survey methods are all presented.

13.1.1 The study region and the database

An area of 17 034 km² including 1534 km² of water was chosen as the study area. The land area was within the King Country region of the central North Island of New Zealand.

The King Country has complex topography and stream patterns. The drainage system serving the ranges dissects the region so thoroughly that north–south road and rail movement is restricted to one main corridor. Access into the countryside by east–west routes often follows river valleys upstream, terminating in the mountain ranges. Extensive tracts of land are remote from all-weather roads or have no access at all. Development has been hindered over large areas by the lack of suitable access. Thus, more than half of the land presently thought of as suitable for agricultural purposes is either undeveloped or only moderately improved.

The image selected for use in this study was acquired by the Multispectral Scanner on board Landsat 2 on 15 February 1976 (GMT)— scene reference number 2389–21 172. During the early months of 1976 aerial photographic surveys were taken to assist field parties in gathering information for a major land use survey of the King Country. Both these factors were important as the resultant thematic mapping product from this remote sensing technology was to be compared with the results of conventional mapping methods for accuracy of classification and registration, time- and cost-effectiveness. Results of this comparison are recorded later in this chapter.

13.1.2 The computing system used in the analysis

The ERMAN II computing package (IBM 1977), installed on an IBM 3033 computer in Sydney, was made available to the project through the New Zealand Govt/IBM Joint Research Program Agreement. For specifics on the package and its operation, see Chapter 9.

13.1.3 Project objectives

The overall objective of this King Country study was to produce a land cover classification map from Landsat data using currently available remote sensing and computer technology. Answers were also sought to the following more specific questions:

(i)   How many classes could be separately identified?
(ii)   How accurately could the image be registered on a standard map grid?
(iii)   What classification accuracies could be achieved?
(iv)   How does the cost and speed of production of a Landsat data-derived classification map compare with a conventional land cover map?

13.2 Analysis Pathway

Chapter 9 describes each step in detail. The steps described below had facets peculiar to this study.

13.2.1 Hue and texturally enhanced Landsat images

This study concentrated on land cover types other than forestry (for the latter, see Chapter 12).

Good photographic imagery was prepared from the Landsat data to select homogeneous training fields for each class (Joyce 1978) and to identify ground control points to enable the image to be registered to a map projection. The images were produced as positive colour transparencies which were then photographically enlarged to 1:100 000.

Road and river detail from New Zealand Mapping Service No 242 (NZMS 242, 1981) sheet 2 was combined photographically with the enhanced unregistered image to a scale of 1:500 000. Although registration between Landsat data and the linework was not exact, the product assisted in locating positions in the field.

13.2.2 Marshalling of existing land cover data

Information was obtained from:

(i) *King Country Land Use Study*, existing land use maps 1977 (NZMS 288, 3 sheets).
(ii) *Aerial Photographs Survey* Nos 2974, 2920, 3838, 5410, 5014, 5147 (1976, 1977, 1979).
(iii) Topographical Maps: NZMS 1, scale = 1:63 360; NZMS 18, scale = 1:250 000.
(iv) *Geological Survey and Soil Maps*, NZ Geological Survey and Soil Bureau (DSIR 1973).

13.2.3 Field visit 1: familiarisation

The terrain and land use practices of the extensive study area were viewed from an aircraft during a reconnaissance trip to establish the classes which were to be discriminated between. Classes and mapping requirements were also discussed with local Department of Lands and Survey staff. Several areas were visited on the ground to see the varied land uses and vegetation types throughout the region.

13.2.4 Class selection

Final class selection was made after considering the enhanced Landsat imagery, ground truth data, the existing land use maps (NZMS 288, 1977) and the Department of Lands and Survey requirements. Fifteen classes were chosen:

High vigour vegetation	Improved pasture, predominantly rye grass and clover species, showing lush growth. Often dairying areas on flat land. Usually cultivated and sown.
Developed rangeland	Improved pasture showing signs of stress due to grazing or moisture deficit. Usually on hill country, frequently top-dressed but rarely cultivated.
Undeveloped rangeland	Unimproved pasture, dominated by native grasses (for example fescue, browntop) usually on hill country, often weed infested. Top-dressed either infrequently or not at all.
Reverting land	Forest areas felled or burnt, left to revert to pasture. Also includes areas dominated by fern and thistle cover.
Wetlands	Swamp and permanently wet areas.
Water	Both salt and fresh water areas.

Ferns	Brackens.
Scrub	Kanuka and manuka dominant.
Exotic forest	Both forest plantations and farm wood-lots.
Native forest	A broad class but mainly native hardwood trees.
Native forest (sunny aspect)	As above but on sunlit faces.
Native podocarp forest	Podocarp species emergent over hardwoods.
Tussock	Flax and tussock (around Desert Road).
Bare ground	Includes exposed rock, cleared forest land and river gravels.
Urban	Cities and towns.

13.2.5 Selection of training fields

Representative areas for each class were needed to 'train' the computing system to produce the required classification map (see Joyce 1978). For each of the fifteen classes it was necessary to select several 'homogeneous' areas distributed throughout the study area. Initial training fields were located on a lineprinter character map of the Landsat image. The field radiance levels in each MSS band were extracted for use in a local parallelepiped classification process before the ERMAN work was begun in Australia. Identifying and marking training fields on aerial photographs aided the analyst in the interpretation of the enhanced imagery and also in the selection of further fields.

Selection of final training fields was accomplished using the combination of enhanced imagery, lineprinter output and aerial photographs. This detailed inspection of the enhanced imagery gave a very clear indication of how the differing radiance levels on the lineprinter output could be combined to give varying densities of colour on the imagery. The imagery used had been texturally enhanced using the Laplacian algorithm and then hue enhanced for pastoral and agricultural areas. MSS bands 4, 5 and 7 were written out through the New Zealand Optronics Colorwrite C-4300 filters: blue, green and red respectively. 'High Vigour Vegetation' areas showed as bright magenta while the 'Developed' class appeared as a flame red, the 'Undeveloped Rangeland' class was seen as a yellow–buff colour. 97 training fields were selected for checking on the next field visit.

The statistics from the selected training fields were used in the trial parallelepiped classification. This trial evaluated the separability of the classes and the homogeneity of the training fields. The four MSS bands were used in this classification and this step helped to define potential problem areas, and assist in their removal, for both class and training field selection.

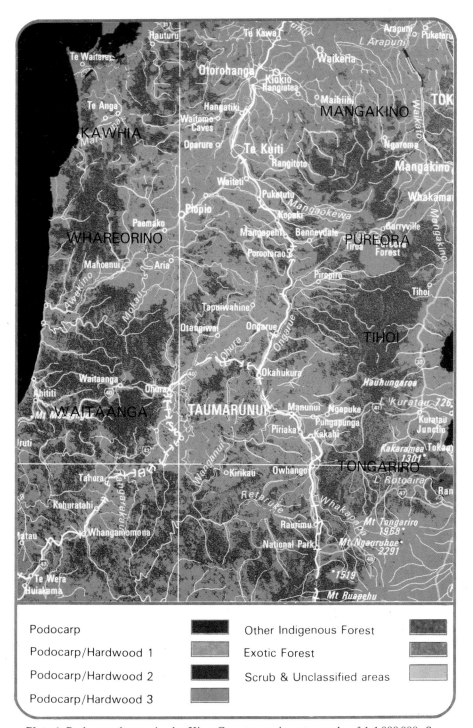

Plate 1 Podocarp forests in the King Country region at a scale of 1:1 000 000. See plate 3 for a 1:63 360 enlargement.

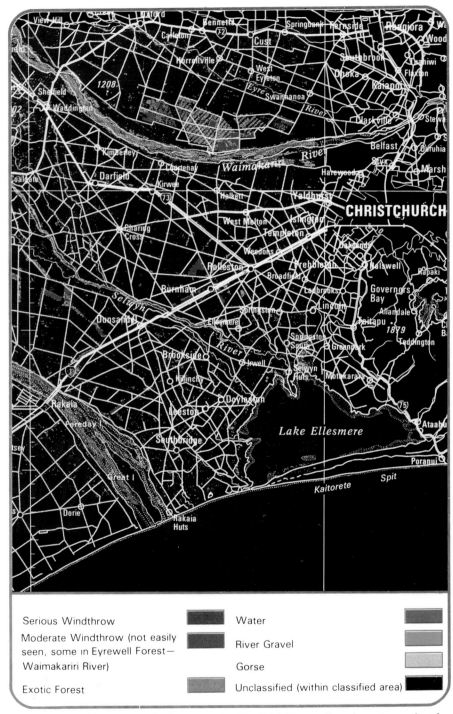

Plate 2 Windthrown exotic forest in the Darfield–Eyrewell region at a scale of 1:500 000. This was a fifteen-channel single-acquisition classification.

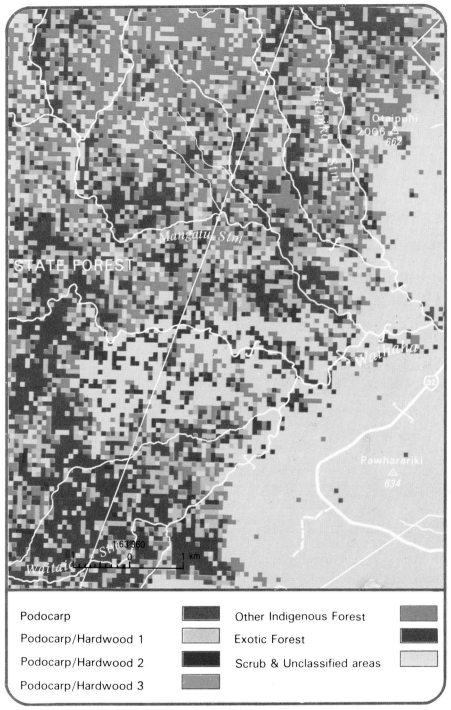

Podocarp	▨	Other Indigenous Forest	▨
Podocarp/Hardwood 1	▢	Exotic Forest	■
Podocarp/Hardwood 2	■	Scrub & Unclassified areas	▢
Podocarp/Hardwood 3	▨		

Plate 3 Podocarp forests: Tihoi state forest in the King Country region at a scale of 1:63 360. See plate 1 for a 1:1 000 000 scale image and figure 12.4 for a forest and scrub map.

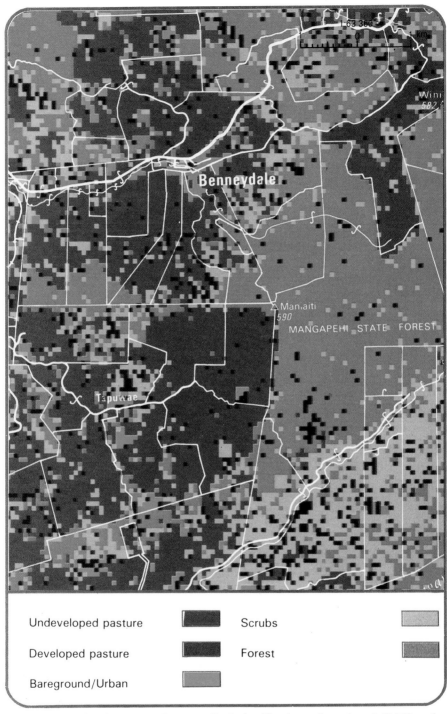

Undeveloped pasture	■	Scrubs	■
Developed pasture	■	Forest	■
Bareground/Urban	■		

Plate 4 KING21CR Image 8, generalised classification enlarged to 1:63 360 and combined photographically with base linework of NZMS 288.

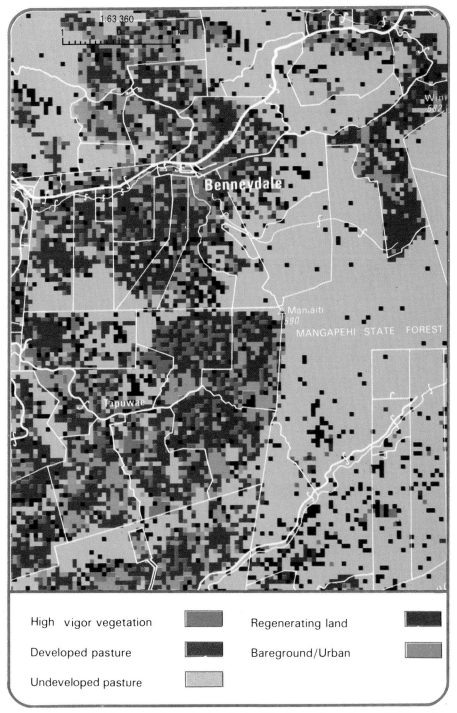

Plate 5 KING21CR Image 1, agricultural classification enlarged to 1:63 360 and combined with linework from Sheet 1 of NZMS 288 King Country Land Use Study.

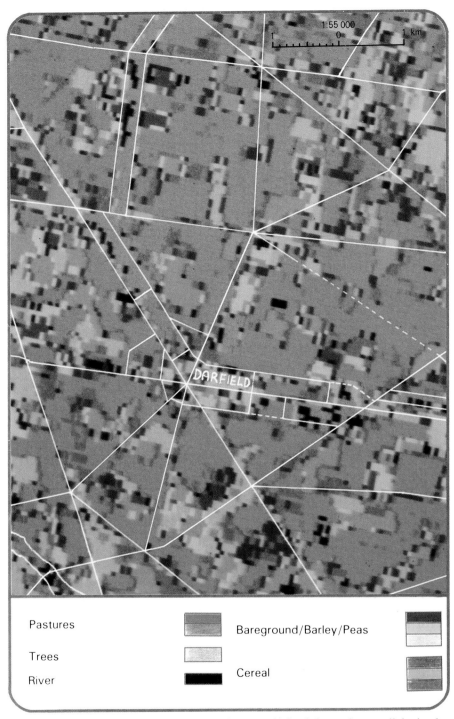

Plate 6 Thematic map of Darfield study area derived from the parallelepiped classifier. Base data: 31 October 1975 Landsat scene 2282–21 254.

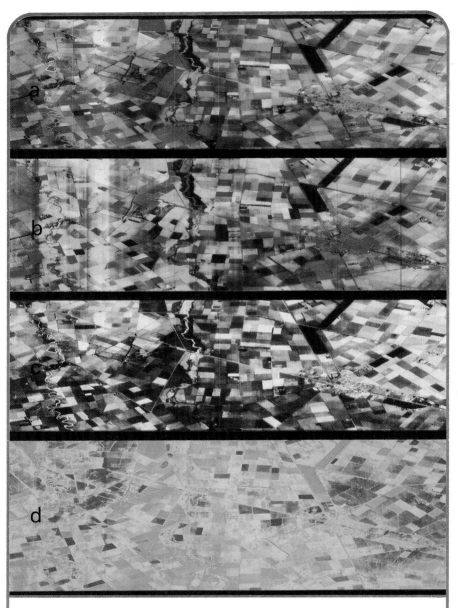

(a) Channels 3, 6, 10—Landsat equivalent channels.

(b) Channels 3, 9, 10—best of three channels for discrimination of Darfield agricultural classes (see chapter 7)

(c) Channels 2, 4, 6—'natural colour' image.

(d) Channel 11—thermal channel.

Plate 7 Corrected aircraft scanner imagery of the Darfield region. Details of the wavelength ranges for each channel and the corrections necessary are given in Chapter 5.

Plate 8 Final Level IV classification map of Darfield agricultural cover derived from eleven-band aircraft scanner data, flown 9 January 1981 — austral summer, using the ERMAN classification package with 0% threshold applied. The scale is 1:60 000. The classification accuracy of the map is 79% with 99.9% confidence limits (see Chapter 11).

13.2.6 Selection of ground control points

Twenty seven points were selected using the enhanced Landsat imagery. Longitude/latitude coordinates, to the nearest second, were scaled for each point from NZMS 1, 1:63 360 topographical maps. Figure 12.2 shows the distribution of GCP.

13.2.7 Field visit 2: training field and GCP checks

The study area was revisited to re-assess the training fields in the light of the image enhancement and the parallelepiped classification. Further, and modified, training fields were selected as a result of this revision. This was only possible because the classes selected (§ 13.2.4) were largely spatially constant from the date of the imagery—as discussed in § 13.2.18. On this visit, GCP which were difficult to find on the Landsat image were well documented and photographed to aid identification on the ERMAN screen.

13.2.8 Computer analysis step 1: registration

The Landsat data were registered to a map base to enable the overlay of linework on top of the final classification product to be achieved. The latitude and longitude values of the selected GCP were entered on the alphanumeric monitor and the locations indicated by the trackball cursor on the image screen. The image was registered to the selected map projection using two iterations, as follows.

Twenty seven GCP and a first-order polynomial were used in the first iteration over an area with the following boundaries:

Most northerly latitude = S 38° 01' 30"
Most southerly latitude = S 39° 20' 00"
Most westerly longitude = E 174° 30' 00"
Most easterly longitude = E 175° 51' 00".

A smaller region was defined for the second iteration, with boundaries:

Most northerly latitude = S 38° 02' 30"
Most southerly latitude = S 39° 19' 00"
Most westerly longitude = E 174° 30' 00"
Most easterly longitude = E 175° 50' 00".

The second iteration was made to improve the accuracy of fit between the first iteration and the NZMS 1 maps. GCP were inserted for this latter step and a first-order polynomial was used again.

For both registration exercises, the data were resampled with an 80 by 80 m^2 pixel size (IBM 1977 and Chapter 8). If a higher resolution had been used (for example a 40 by 40 m^2 resampling pixel size), the resulting data volume and computing time would have been excessive for this project.

To check the registered images quickly, the images were enlarged to approximately 1:250 000 at the computer terminal and a map transparency at that scale placed over the screen. A detailed check of registration accuracy was conducted later (§ 13.2.14).

13.2.9 Computer analysis step 2: synthesis of extra channels

The four-band registered data was used as the base data. Further channels were created from the four MSS bands: four Principal Components and twelve Band Ratios (see Chapter 5 for full details on the derivation of these synthetic channels). Channel allocation in the new twenty-channel image is shown in table 13.1.

Table 13.1 Channel allocation in twenty-channel image.

Channel	Data	Channel	Data
1	PC1†	11	'5/6'
2	PC2	12	'5/7'
3	PC3	13	MSS 6
4	PC4	14	'6/4'
5	MSS 4	15	'6/5'
6	'4/5'‡	16	'6/7'
7	'4/6'	17	MSS 7
8	'4/7'	18	'7/4'
9	MSS 5	19	'7/5'
10	'5/4'	20	'7/6'

† Principal Component.
‡ A ratio of MSS band 4 to MSS band 5 (following the algorithm detailed for ERMAN in Chapter 6).

13.2.10 Computer analysis step 3: insertion of training fields

As the training fields form the basis of the whole classification this step was a most exacting task.

A suitable image enhancement for training field identification for this land cover project was selected:

Channel 13   MSS 6 through red colour gun
Channel 5   MSS 5 through green
Channel 1   PC1 through blue.

Training fields were carefully identified and a polygon with a (system) maximum of nine independent vertices drawn around the field using the trackball and cursor. Since each vertex had to be separated from its neighbour by at least one pixel width, very small training fields could not be used.

The next stage was to enlarge each training field to inspect the spectral variation (seen as colour variation) within each field. The boundaries of training fields with too much or too little spectral variation were revised (Chapter 6).

Each training field was identified by a Field Name, Field Class Name and Symbol. The Symbol was used to represent the class in the final output character maps and Divergence reports. To satisfy the 'n + 1' minimum pixel number criterion (see Swain and Davis 1978) at least 21 pixels were needed in the training set for each class. Initially 137 training fields were identified to represent 15 classes but the number of fields was significantly reduced after statistical analysis.

13.2.11 Computer analysis step 4: compilation of statistics

The training field statistics were collated into histograms and viewed. Mean vector and variance/covariance matrices were computed from the training field data. Standard deviations were also computed. Mean and standard deviations were then assessed.

Training fields which had standard deviations that were too high were amended, relocated or deleted. After revision, 39 training fields representing the 15 classes remained for final classification (§ 13.2.4).

13.2.12 Computer analysis step 5: channel selection via Divergence

Divergence enables the user to select the optimum set of channels for classification (see Chapter 7).

Divergence, run in sets of eight channels over all classes, ranked the channels according to the average Divergence values over all class pairs (table 13.2). Those class pairs with 'poor' separability, from the full D_{ave} report, were then investigated further with the Divergence module and the best set of four channels to separate each weak class pair was found (table 13.3). (The same criterion was used to define 'weak class pairs' as was used in § 12.14.15.)

Table 13.2 Best eight channels ranked by average separability over all class pairs.

Channel	Data	Channel	Data
5	MSS 4	12	'5/7'
7	'4/6'	13	MSS 6
8	'4/7'	16	'6/7'
9	MSS 5	17	MSS 7

Table 13.3 Additional four channels needed to separate 'weak' class pairs.

Channel	Data	Interclass pairs
1	PC1	High vigour–developed rangeland
		Exotic forest–sunlit forest
2	PC2	Native forest–sunlit forest
		Native forest–exotic forest
11	'5/7'	High vigour–developed rangeland
		Native forest–sunlit forest
20	'7/6'	High vigour–developed rangeland
		Scrub–tussock
		Native forest–sunlit forest

13.2.13 Computer analysis step 6: test classifications of a sample area

Test classifications were prepared for field checking/assessment in New Zealand before executing the classification of the full region.

13.2.13.1 Test Classifications

Two initial classifications of the selected training field data were made. The first used the best eight channels from the ranking given by D_{ave}. The second used the best twelve channels from both D_{ave} and from D_{ij}. To reduce computing a small test field was chosen as a trial classification area. After each classification run on this trial, statistical reports were compiled and displayed to show the number of pixels in each class.

Several tests were carried out to investigate the variations in statistics using different thresholds for the eight- and twelve-channel classifications (table 13.4). The variations produced by different choices of threshold values were also recorded for the twelve-channel classification from the class summary reports (table 13.5). Each classification was displayed on the colour screen and also recorded as a lineprinter character map.

Table 13.4 Number of pixels classified in four different classes using different 12- and 8-channel sets and two different threshold values.

	Classes			
	Reverting	High vigour	Developed	Scrub
8-channel 0% threshold	38 930	8366	18 184	22 459
12-channel 0% threshold	39 421	10 430	17 033	20 563
12-channel 1% threshold	35 819	9473	16 589	18 718

Table 13.5 Class summary using twelve-channel classification and differing threshold values.

Symbol	Class	Pixels in training	Thres 0%†	Thres 1%†	Thres 3%†	Thres 10%†
R	Reverting land	105	93.3	92.4	92.4	85.7
V	High vigour veg.	83	89.2	88.0	88.0	86.7
U	Urban	115	96.5	93.9	91.3	87.0
B	Bare ground	105	94.3	93.3	93.3	86.7
D	Developed rangeland	40	92.5	92.5	92.5	90.0
G	Undeveloped rangeland	157	95.5	94.9	94.3	90.4
M	Wetlands	21	100.0	100.0	100.0	81.0
H	Water	3463	100.0	96.5	96.5	94.4
K	Scrub	45	80.0	80.0	77.8	75.6
F	Fern	55	98.2	96.4	92.7	89.1
X	Exotic forest	102	93.1	92.2	91.2	87.3
S	Forest	98	78.6	77.6	77.6	74.5
L	Forest (sunny aspect)	126	80.2	79.4	78.6	77.8
I	Indigenous forest	202	93.1	92.1	89.6	88.1
C	Tussock	236	87.3	86.0	85.6	81.8

† Percentage of pixels in training fields actually classified as belonging to the nominated class for the field.

13.2.13.2 *Preparation of coloured and character maps*

Character maps of the test area were generated, for the eight- and the twelve-channel classifications, with both 0% and 1% thresholds. These maps were prepared using modified software (Beach 1980) on the IBM 3800 laser printer at a scale of 1:37 800.

Careful planning was needed at this stage to produce colour class maps since the colour screen provided only six colours, excluding black and white. Eight-colour class maps were also displayed to show the differences between the eight-channel and the twelve-channel classifications and also the effects of different thresholds.

Since the agricultural classes were of prime interest in this study, the scrub and forest classes were all displayed as white (see plate 5). The colours allotted to classes were as follows.

Colour No	Colour	Classes
0	Black	(Thresholded areas) unclassified areas
2	Blue	Water
4	Green	High vigour
6	Cyan	Bare ground and urban

Colour No	Colour	Classes
8	Red	Developed
10	Magenta	Reverting land
12	Yellow	Undeveloped
14	White	Scrub, fern, tussock, all forest classes

The eight-colour images were combined after viewing into a multiband classification image and written to tape for subsequent transcription to colour film in New Zealand on the Optronics Colorwrite.

13.2.13.3 Field visit 3: check of test classification area

The character maps of the areas for field checking were hand-coloured to highlight the different classes and to aid the location of roads. The latter was important to effective field checking. Colour class maps similar to those displayed on the ERMAN screen were produced on the Colorwrite to support this evaluation.

The accuracy of classification of the pixels in the test region was recorded as correct or incorrect. Ground checking covered 0.15% of the total test field—409 pixels.

The method described in Chapter 11 was used to determine the accuracy of the classification. Over the limited area, for the classes tested, the probability of a pixel being classified into its correct class 977 times out of 1000 was 80.7%. This was considered to be sufficiently accurate for thematic mapping of this type and gave confidence in the classification technique used.

The only major modification was the combining of the urban and the bare ground classes. The spectral signatures of these classes were almost identical and therefore difficult to separate.

13.2.14 Check of registration accuracy

The map projection used in the ERMAN computer system is a non-conformal projection in which the orthogonal coordinate system is defined by lines of longitude and latitude, while the maps used to overlay road/river information on the registered image for New Zealand are on a Transverse Mercator projection (NZTM) with its North Island origin at 175° 30′ E longitude, 39° 00′ S latitude.

A comparison was made between the registered image and the New Zealand Mapping Service 1 (NZMS 1) topographical maps to determine the overall accuracy. This overall accuracy is considered to be a combination of a number of factors.

(i)   Identification of ground control points (GCP).

(ii)   Reading error of latitude and longitude values for GCP.
(iii) Locating GCP on ERMAN monitor.
(iv)   ERMAN registration.
(v)   Difference between ERMAN and NZTM projections.
(vi)   Identification of the assessment points on ERMAN product and NZMS 1.
(vii) Reading error of NZTM coordinate values for assessment points.

Thirty five well-defined assessment points were measured (X, Y) on the registered Landsat image from ERMAN using a coordinatograph. From the NZMS 1, their corresponding map coordinates (E, N) were recorded in terms of the New Zealand National Yard Grid.

Using the registered image coordinates (X, Y) and the map coordinates (E, N), polynomials were calculated. The residuals, or deviations, between the transformed values and the map values were then determined.

The standard error over 35 points was calculated giving an overall result of 125 m. This represents an accuracy of approximately 1.0 mm at 1:100 000 scale, 2.0 mm at 1:50 000 (1.5 Landsat 80 m pixels).

The overall registration accuracy of $\pm$ 125 m is considered to be acceptable for thematic mapping at scales of 1:50 000 and smaller.

13.2.15 *Computer analysis step 7: classification of the full desired region*

As the test classification results proved satisfactory the same class statistics (see §§ 13.2.11, 13.2.12, and 13.2.13) were used to classify the full region. This region comprised 2.6 million pixels and these were all individually classified into one of the fifteen classes.

Field and class summaries were reported, viewed and copies taken. The classification dataset (see Chapter 7) was also recorded and written out to tape for subsequent use in New Zealand.

13.2.16 *Computer analysis step 8: preparation of final character maps and colour images*

The 1% threshold was applied to all classes and copies of the lineprinter character map were produced.

Colour images of the classification were also produced, each concentrating on a different theme of the classified data. Nine different classification images were created and stored on one tape as a multiband image. Generally, water was depicted as blue, thresholded out areas were shown in black and the combined bare ground/urban were displayed in cyan, as this corresponded well with a Landsat MSS 4, 5, 7 image. (Blue and cyan colours were easily differentiated on the screen and for this reason water and urban areas were easily distinguished in each of the colour images, a useful location feature.)

The colour codes for the full nine classification images, being mindful of the fifteen classes sought in § 13.2.4, are presented in table 13.6.

'Positive masks' of each class were also prepared by taking each class individually and displaying it in black whilst all other classes were shown in white. These single-colour class images could be enlarged and combined to produce a colour map via the lithographic printing process.

Upon return to New Zealand, the tapes containing the multiband colour and the black and white images were reformatted and written out via the Colorwrite with a 25 micron spot size. A two times expansion was used giving a transparency scale of 1:1 606 000. A program called FORMAT (Burden and Whitcombe 1980) was used to produce exactly the same colours on the transparency as were viewed on the ERMAN screen.

Table 13.6 Colour combinations for Classification Images.

KING21CR	Image 1	Image 2	Image 3
Title	Agriculture	Undeveloped	Bush/scrub
ERMAN No	MAC01MCOLR	MAC02MCOLR	MAC03MCOLR
Black	1% threshold	1% threshold	1% threshold
Yellow	Undeveloped	Fern	Tussock
Magenta	Reverting	Reverting	
Red	Developed	Undeveloped	Fern and scrub
Cyan	Bare ground/ urban	Bare ground/ urban	
Green	High vigour	Scrub	All forestry classes
Blue	Water	Water	Water
White	All other classes	All other classes	All other classes

KING21CR	Image 4	Image 5	Image 6
Title	Water	Forests	High vigour
ERMAN No	MAC04MCOLR	MAC05MCOLR	MAC06MCOLR
Black	1% threshold	1% threshold	1% threshold
Yellow	Wetlands	Tussock	Undeveloped
Magenta		Exotic	Fern
Red		Indigenous	Developed
Cyan	Bare ground/ urban	Forest	Wetlands
Green		Forest– sunlit	High vigour
Blue	Water	Water	Water
White	All other classes	All other class	All other classes

KING21CR	Image 7	Image 8	Image 9
Title	Farm woodlots	Generalised	No threshold
ERMAN No	MAC12MCOLR	MAC13MCOLR	MAC00MCOLR
Black	1% threshold	1% threshold	0% threshold
Yellow	Fern/scrub	Tussock/ fern/scrub	Undeveloped
Magenta	Undeveloped	Undeveloped/ reverting	Reverting
Red	Exotic	Developed/ high vigour	Developed
Cyan	Bare ground/ urban	Bare ground/ urban	Bare ground/ urban
Green	High vigour/ developed	Forest classes	High vigour
Blue	Water	Water	Water
White	All other classes	All other classes	All other classes

Colour prints of the nine class combinations for the whole test area were produced at 1:1 000 000. Areas to be studied in greater detail were further enlarged to scales of 1:250 000, 1:100 000 and 1:63 360 to compare with existing King Country Land Use Maps (NZMS 288) (plates 4, 5 and figure 13.1). These scales were chosen to correspond to existing topographic maps to aid location in the analysis and field checking exercises.

13.2.17 Linework overlays for classification maps

Linework from several topographic mapping series was used to combine with colour classification images. To display the whole of the classified area on the one print, a 1:250 000 scale was selected. Road, typographical and hydrographical detail from the NZMS 18 series was combined to a transparent film and a negative produced of the required area. A white topographic overlay was produced photographically on the classification map; the latter enlarged to fit the linework (Clode 1980).

Subscenes within the test field (§ 13.2.13) were also enlarged to 1:63 360 for combination with base linework from Sheet 1 of NZMS 288 King Country Land Use Study. The resultant product was useful as a comparison with the existing Land Use Sheet (plates 4, 5 and figure 13.1).

The complete area of the classification had also been recorded in lineprinter form and this character map was joined together and then cut into sheets to correspond with the NZMS 1 sheet lines. To establish where sheet lines fell on the character map, it was necessary to locate points in each corner of the 15 NZMS 1 sheets. These points were located on the image displayed on the ERMAN colour TV screen. Using the Image Manipulation and Display module, line and pixel values as well as latitude and longitude values were displayed on the alphanumeric screen for each point. These line, pixel and latitude and longitude values were used to establish a transformation formula to find the line and pixel values for each of the corners of the NZMS 1 sheets. These corner points were located on the character map, as were all GCP to act as a check of the enlargement factor for the overlaid linework.

The character map was measured and its scale established at 1:37 800. At this stage linework information, for the 15 NZMS 1 sheets covering the region, was enlarged 1.68 times to match the scale of the character map. These enlarged transparencies were placed on a light table with the character map sheets over the top. By fitting corner points and GCP, road and river details were traced onto the character maps. This information was found to be extremely valuable for field checking (see Chapter 9).

13.2.18 Field visit 4: overall classification accuracy

A field programme similar to that described under § 13.2.13.3 was conducted to assess the accuracy of classification over the whole region classified. The field check was made in February, the same month as the original Landsat data were collected. The stage of plant growth and moisture conditions were similar to those existing when the imagery was collected five years previously. Forestry and farming patterns had changed very little over the five years although some discrepancies were seen. For example several areas which were classified as 'scrub' had been cut and burnt and by February 1981 were rough pasture.

Taking the above factors and grazing patterns into consideration, eleven classes were selected for checking over twelve regions within the classified area. (Bare ground and urban were checked as one class and the four forestry classes were also considered together.) Twelve regions were distributed throughout the entire study area to give maximum representation of different soil types, climatic conditions, terrain and vegetation communities. The numbers of pixels correct and incorrect were tabulated by class.

This extensive field checking programme yielded very encouraging results on a class-by-class assessment. As indicated in table 13.7 the lowest confidence level was 82.7% in the 'wetlands' class. This class had only 1883 pixels over the whole classification and many of them were inaccessible. It is possible that had a greater number of pixels been checked, the confidence level may have improved.

Results were tabulated on a regional basis and no significant variation between regions was seen. Of the total of 2 661 552 pixels classified, 0.97% were checked and 24 587 (out of a total of 25 773) pixels were found to be correctly classified. Over all the classes the probability of a pixel being classified into its correct class 999 times out of 1000 is 94.5% ± 0.5% (based on the results of table 13.7). (For a more complete discussion of the reduction of the analysis see the example used in Chapter 11.)

Table 13.7 Probability of a pixel being classified correctly 999 times out of 1000.

Class	No pixels checked n	No correct p	No incorrect q	Binomial st. dev. $(npq/n^2)^{1/2}$	99.9% CL $100(p-3s)/n$
High vigour	1618	1441	177	12.5	86.8
Developed	2250	2054	196	13.3	89.6
Undeveloped	2558	2449	109	10.2	94.6
Reverting	5190	4960	230	14.8	94.7
Ferns	2526	2426	100	9.8	94.8
Scrubs	2453	2379	74	8.5	96.0
Forests	3308	3241	67	8.1	97.2
Tussock	1695	1661	34	5.7	96.9
Bare ground/urban	3173	2977	196	13.5	92.5
Wetlands	52	49	3	1.6	82.7
Water	950	950	0	—	100

13.2.19 Output products
For a more complete report on these products, please refer to Chapter 10.

Colour maps. Colour classification maps of many of the combinations described above were produced as a photographic print with road and river detail combined. The scales of the prints differed to suit different requirements but generally 1:100000 was the preferred scale (§ 13.2.17).

Character maps. The character lineprinter output and a cartographically produced road/river overlay (NZMS 1) were both photographically reproduced at 1:50000 and combined. Using a Xerox 2080 (a scale-changing copier) a transparent film copy was done to allow relatively inexpensive copies to be made (§ 13.2.17). An example of this is seen in figure 12.3.

13.2.20 Cost comparison
In 1977 part of the King Country Region was surveyed and mapped by conventional methods to produce the King Country Land Use Study.

Only the costs of gathering the information and drawing a visual record (at 1980 values) have been considered in tables 13.8 and 13.9; the printing cost has been excluded.

Table 13.8 Cost analysis of conventional mapping methods. King Country conventional survey—area 4376 km².

	1978 (NZ$)	1980 (NZ$)
Field work, travel, mileage, research, data collection	16 500	22 300
Aerial photography	26 800	42 880
Mapping	16 100	25 800
Servicing and processing	2100	3360
	61 500	94 340

Table 13.9 Cost analysis of Landsat/ERMAN mapping methods. King Country Landsat/ERMAN survey—area 17 034 km² (includes 1534 km² of water).

	1980 (NZ $)
Staff, 1 man-year,	17 000
field, office	800
Travel and vehicle costs	3200
Stores (maps, prints, CCT US$200)	3290
Computer support system	6000
ERMAN 2.4 CPU hrs	est 10 000
(cost estimate based on NZ Ministry of Works and Development, Vogel Computer Centre charges)	
Travel (Australia) 3 trips	1120
Allowance (Australia) 39 days	2925
	44 335

The cost of producing a land use map by conventional survey is NZ$21.56 per km²; the Landsat/ERMAN survey was NZ$2.86 per km². Furthermore, the Landsat/ERMAN method surveyed 15 500 km² per man-year whereas the King Country Land Use Study was achieved at 4376 km² per man-year. Thus a fiscal saving of 87% and a manpower saving of 72% was made by the ERMAN survey over the King Country Land Use Study.

13.3 Comparison of Approaches

Mr W A Robertson, the Director of Planning for the New Zealand Department of Lands and Survey (1980) compared the Landsat/ERMAN map with the King Country Land Use map, and concluded:

'There is very good agreement on the broad categories of land (cover) which have been derived from remote sensing and manually. Generally, the boundaries of categories correspond closely.

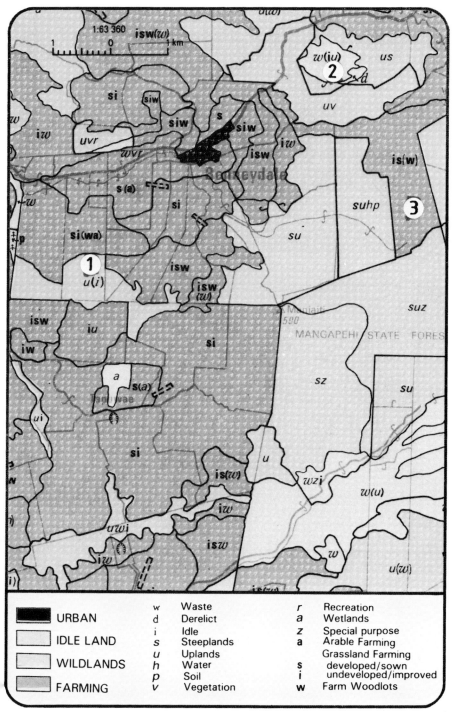

Figure 13.1 Black and white representation of Sheet 1 of NZMS 288 King Country Land Use Study showing existing land use in 1977.

In addition the remote sensing maps provide very detailed patterns of information to a degree not practical with manual methods. The following examples provide specific comparisons between three areas numbered 1 to 3 on figure 13.1.

(**1**) The light grey area coded $u(i)$ on the land use map denotes an upland bush area with a small percentage of idle land.

The remote sensing map confirms the forest outline and indicates the *position* of small scrub areas or idle land.

(**2**) The lightest grey area coded $w(iu)$ and d on the land use map denotes wasteland with a small portion of partly improved land. The area coded d indicates derelict land.

The remote sensing map indicates that the wasteland is scrub and it pinpoints the small areas of improvement pasture (red). The derelict area is identified as bare ground.

(**3**) The darkish grey area coded as (**w**) on the land use map denotes a major undeveloped grass area with a significant component of developed grassland and a small percentage of woodlots.

The remote sensing map indicates that this is correct but the fine detail shows the actual area and location of each of these components. The developed areas in the north of the unit and the actual site of the small forest lots are evident. The balance area, which is the major area, shows up as undeveloped grassland.

The results achieved are positive and indicate that remote sensing offers a quick means of obtaining land (cover) in very detailed patterns. An overlay of land use data and statistics for each holding would provide accurate and comprehensive general land use information.'

13.4 Conclusions

There is an increasing need for more careful and frequent monitoring of our natural resources, but monitoring costs are rising and staff numbers diminishing. The Landsat/computer analysis system can therefore make a useful, cost- and time-effective, and sufficiently accurate contribution to the process of assessing and managing resources.

The applicability of an advanced statistical system for the preparation of a land cover map has been tested for a part of New Zealand. The area was difficult to map using conventional methods as access to this region was difficult. Although the terrain ranges from flat pumice land to deeply dissected mudstone it supports highly spatially diverse land use practices. These factors have all helped to test the suitability of computer mapping techniques. The 94.5% confidence level for the classification is acceptable for future land inventory mapping as is the ± 125 m accuracy for registration with relation to the ground.

Classes chosen for the classification were very successful although more detail could have been extracted from the data had the classes been more closely defined. With the promise of greater resolution data from future spacecraft, the introduction of larger computing systems and the increased availability of remotely sensed data from space, this method of classification could provide a valuable tool to those managing and planning the future use of a country's resources.

13.5 References

Aerial Photographs Survey Nos 2974, 2920, 3838: 1976; 5014: 1977; 5147, 5410: 1979 (Hastings, NZ: New Zealand Aerial Mapping)

Beach D W 1980 Informal communication, IBM, Sydney

Burden A K and Whitcombe A N R 1980 *FORMAT—a Program Used to Edit the ERMAN Series of Magnetic Tape on the Varian Computer* Report No 669 (New Zealand: Physics and Engineering Laboratory, DSIR)

Clode D 1980 Informal communication, DAC Productions, Wellington

DSIR 1973 *Geological Survey and Soil Maps* (Wellington, NZ: NZ Geological Survey and Soil Bureau)

IBM 1977 *Earth Resources Management II (ERMAN II) User's Guide* Program No 5790-ARB Doc. No SB 11-5008-0 (New York: IBM)

Joyce A T 1978 *Procedures for Gathering Ground Truth Information for a Supervised Approach to a Computer Implemented Land Cover Classification of Landsat-Acquired Multispectral Scanner data* NASA Ref. Publ. 1015 (Washington, DC: NASA)

NZMS 1 various dates 1971–82 (Wellington, NZ: Department of Lands and Survey)

NZMS 18 various dates 1964–75 (Wellington, NZ: Department of Lands and Survey)

NZMS 242 1981 (Wellington, NZ: Department of Lands and Survey)

NZMS 288 1977 (Wellington, NZ: Department of Lands and Survey)

Swain P H and Davis S M 1978 (ed.) *Remote Sensing: The Quantitative Approach* (New York: McGraw-Hill)

14 | Classification of Agricultural Land Cover from Landsat

14.1 Introduction

New Zealand is principally a primary producing country (New Zealand Yearbook 1981) and pasture and cereal farming play a large role in the nation's economy.

Some parts of the growing cycle are more suitable than others for a satellite to discriminate between crop types. For instance, in New Zealand, moisture stress only occurs at certain times of the year.

The largest region of arable farming in New Zealand is in the Canterbury Plains, which border the east coast of the South Island. The terrain is very flat. This avoided the need to consider the shading effect of sloping ground, which can itself confuse target identification. The type of farming is very diverse. The practice is to rotate the land between pasture and arable crops—with legumes used at intervals in order to raise the nitrogen levels in the soil—and for forage or seed production.

The test site chosen was around Darfield, a small agricultural town inland from Christchurch. With the cooperation of farmers in the Darfield region, an inventory of the area was conducted which covered about 400 fields. The average field size was 9.6 hectares. At the resolution of the Landsat (1, 2, 3) Multispectral Scanner this represented 20 pixels per field.

In this chapter, the results of classifying the data using the simple parallelepiped method as well as the more sophisticated Maximum Likelihood system (see Chapter 7) are presented.

This study had the following objectives:

(i)   To determine the levels of classification (see Chapter 3) which could be attained from Landsat images. In particular it was desired to separate the wheat class from any other cereal classes.

(ii)   To compare the results of the two classifiers (parallelepiped and Maximum Likelihood).

(iii)   To compare the results of classifying with one image with the results obtained using two images acquired at different times of the year. The location of the study area is shown in figure 14.1.

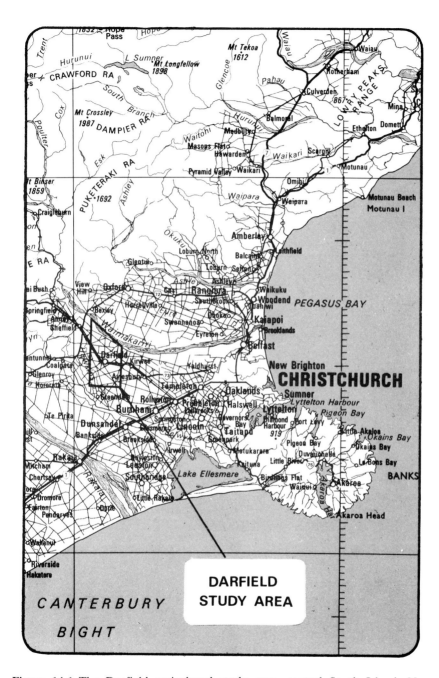

Figure 14.1 The Darfield agricultural study area, central South Island, New Zealand.

14.2 The Study Area Cropping Programme and Data Acquisition

The cultivation practices for the major crops in the Darfield area are summarised in table 14.1. From the table we can deduce that there are two periods during the year when farm conditions conducive to classification occur: June through to August, before spring planting starts; and December, before harvest. These two periods are probably best for comparing temporal changes in the spectral signatures of each crop type and should distinguish autumn-sown wheat from the other cereal crops. However since most of the crops except lucerne (alfalfa) and pasture grass are sown in the spring, late October and late December imagery of the same growing season are probably the best to use for temporal comparisons. (The farming practice is to rotate the lucerne and grass fields and sow them in crop every three to ten years, depending on the soil type.)

If the crops are to be classified from one coverage only, table 14.1 would suggest that December is the optimum time.

Table 14.1 Typical Crop Management Schedule for Darfield, Central Canterbury, New Zealand. The key is given below.

CROP	JUL	AUG	SEP	OCT	NOV	DEC	JAN	FEB	MAR	APR	MAY	JUN
AUTUMN WHEAT	▷		◁ o					◀S	SS	S B	B▶	B B
SPRING WHEAT	B B	▶B	B B	B				◀S	S S	S S	SB	B B
BARLEY	B B	B B	B B	B B	B			◀S	SS	SS	S B	B B
OATS	B B	B▶	B B					◀S	S S	SS	S B	BB
SPRING PEAS	B B	B B	B▶	B B			◀	BB	BB	B B	B B	BB
AUTUMN PEAS								◀B	BB	▶B	B B	BB
RAPE SEED							◀	BB	▶B	B		
RAPE FEED		B B	B▶	B B	▷				◁B	BB		
LUPIN	▷		◁			B	BB	▶B	B			
POTATOES	B B	B B	B B	▶B	B B				B	BB	◀ B	B B	BB
SEED KALE					▶		◀B	BB				
LUCERNE			▶B		●● ●●	●●				◀B	BB	BB
CLOVER	▷		▶ B			◀						◁
GRASS	B B	▷					◀	▶ o	B▶	B		◁

KEY		
	PLANTING	▶
	CUTTING	●
	SEED HARVEST	◀
	GRAZING (BEGIN → END)	▷ ◁
	BARE SOIL	B B B B
	STUBBLE	S S S S
	UNDERSOWN WITH CLOVER	o

Except for potatoes, the crops are all planted and managed in much the same manner. The operations in sequence are:

  (i)   ploughing;
 (ii)   cultivating;
(iii)   drilling (with harrow trailing the drill);
 (iv)   rolling when crops are 3″ high (only for wheat and other crops that may have to be cut near the base of the stem when harvesting);
  (v)   undersowing wheat with clover (depending upon the planned rotation);
 (vi)   applying fertiliser;
(vii)   spraying to kill the grass or weeds (varies, but usually done if clover seed is to be harvested);
(viii) harvesting.

Occasionally the autumn wheat and grass crops are undersown with clover. Consequently the spectral signatures of autumn wheat and grass will vary in the spring imagery depending upon the presence and stage of growth of any of the clover. Similarly the spectral signatures of the cereal crops will also vary depending upon the percentage of soil exposed between the plants early in the season. The spectral signatures of all crops will differ later in the season as they reach maturity.

In addition to the usual variations of soil type, soil moisture and microclimate differences, farming practices will also differ within a region. Within each field these influences are, however, often homogeneous. These variations can be allowed for in the classification process by assigning a separate class to each variant of a particular crop type, and then amalgamating the results at the map production stage. This process can work well for major differences such as differing soil types, but becomes more complex in the case of differing farming practices.

Spectral differences within fields are usually due to the cultivation and drilling techniques employed, or to differential grazing or weed infestation. For instance, infestations of wild oats in cereal crops can form a canopy over the crop, and influence the spectral signature. Again, once recognised, this variant can be treated as a separate (say, wild oat/cereal) class and amalgamated in the map production process.

Using high resolution panchromatic aerial photography, obtained in October 1975, the uniformity of all fields within the Darfield study area was examined, using stereo pairs of photographs at a scale of 1:24000. The objective was to assess the possible effects of these irregularities on the spectral signature. The results are summarised in table 14.2.

The spectral signature can also be influenced by edge effects. In some cases a field may be represented by only 15–20 Landsat data pixels. Pixels which cover field boundaries can be influenced by two entirely different spectral signatures, and the result can be an intermediate signature which may be misclassified into a completely different class. Both the field size

Table 14.2 Some influences of farming practice on the homogeneity of class training fields, as observed in aerial photographs from the October 1975 (GMT) Landsat Imagery.

Crop	Grey scale	Field patterns	Farm machinery ploughing	Marks drilling	Grazing indicators
Autumn wheat	Med*	Striated†‡	Yes	Yes	No
Spring wheat	Med	Striated†‡	Yes	Yes	No
Oats	Med	Striated‡	Yes	Yes	No
Rape	Med	Uniform	No	Yes	No
Lupin					Yes
Lucerne	Med+	Uniform	No¶	No	Yes
Clover	Dark§	Uniform	No¶	Faint	Yes
Grass	Med§	Uniform	No¶	No	Yes
Bare soil‖	Light	Uniform*	No	No	No

†May be left unharvested over winter and appear patchy due to storm damage.
‡Depending on the relative growth progression of each class between farmers, the planting/drilling rotation may be more or less evident.
§Most fields are grass/clover or wheat stubble/clover mixtures, with darker tones indicating greater proportions of clover.
‖Fields could be planted in:
barley, peas, potatoes, beets or other root crops.
¶May show faint turning marks if recently planted.
*May show moisture patterns after recent rainfall.
+Varies with condition due to grazing.

and aspect ratio (ratio of length to width) will determine the magnitude of the edge effects. Table 14.3 gives a summary of the measurements of field size and aspect ratio derived from the 96 training fields used in the Darfield Landsat Maximum Likelihood classification.

Table 14.3 Average field size and aspect ratio for fields in the Darfield test area.

Number of fields measured	96
Average number of Landsat pixels per field	20
Average number of resampled 40 × 40 m² pixels per field	60
Average field area (hectares)	9.6
Average aspect ratio	0.654
Standard deviation of the aspect ratio	0.164
Average percentage of resampled pixels which can be influenced by edge effects	46%

Because of the similar crop types in many neighbouring fields, the influence of adjacent fields may not be as important as might be inferred from the last entry in table 14.3. However, as will be seen later, these effects do reduce the accuracy of a classification if the classification has been made purely on a pixel counting basis.

14.3 Summary of Test Site Data

Two images were used for the Landsat database, one from August (winter) of 1975, and the other from October (spring) of the same year.

Two weeks prior to the October coverage, an aerial survey was flown in which high resolution stereo pairs of panchromatic photographs were obtained.

At the time of the winter Landsat overflight (2 August), an aircraft survey was made of the study area using the four-camera multispectral system developed at the Physics and Engineering Laboratory, fitted with simulated Landsat filters (Ellis *et al* 1978). A similar multispectral survey was made within two weeks of the October overflight.

In both August and October 1975, the farms in the Darfield area were visited to obtain detailed ground truth for over 200 fields. The data included crop type, species, height, stress, farming practice and, where appropriate, grazing patterns. The October ground truth survey was followed up by a retrospective visit to resolve apparent anomalies seen on the aircraft and satellite imagery.

Also at the time of the October overflight, measurements were made at the study area of the atmospheric extinction coefficient in each Landsat band throughout the day. This coefficient was used to evaluate the need to correct the Landsat images for atmospheric attenuation and scattering. Table 14.4 lists the images, data and dates collected for the Darfield study area in 1975.

Table 14.4 Darfield test site acquired during 1975.

Data description	GMT date
Landsat scene No 2192–21265	2 August
Landsat scene No 2282–21254	31 October
High resolution panchromatic aerial survey No SN2860	15 October
Aircraft multispectral survey No MSH 4	2 August
Aircraft multispectral survey No MSH 5	14 November
Winter ground truth survey	2 August
Spring ground truth survey	30 October
Atmospheric extinction coefficient data	30 October

14.4 Expected Class Discriminations

The discussion in § 14.3 and the data of table 14.1 give some indication of the expected level of class discrimination.

In Chapter 3 the levels of class discrimination proposed were presented (following Anderson *et al* 1972). These classes were extended to suit the Darfield agricultural study (see table 3.2). The desired classification levels using the two images (August and October) correspond to those listed under Level IV (table 3.2). Moreover, it is hoped that Level V (table 3.2) might be achieved in the cases of spring and autumn wheat, and of standing and mown lucerne. Such a detailed level of classification was not expected using the October scene alone. The prime objective in studies using this single acquisition was to see if cereals, particularly wheats, could be identified and to what level major classes such as pasture could be subdivided.

14.5 Parallelepiped Classification

14.5.1 Method

Only the October image was used for this classification. The aim was to achieve approximately a Level III (table 3.2) classification for the classes listed in table 14.5.

The following procedure was used to define the four MSS band upper and lower radiance limits and to enter these into the parallelepiped classifier (Thomas 1979, see Chapter 6).

(i)   Fields with known ground truth and lying within the Darfield study area were identified on the MSS 5 satellite image. Their boundaries were determined using a transparent overlay map scaled down to fit the image.

(ii)   It was necessary to determine a representative spectral signature for each crop type. Where possible, large spectrally homogeneous fields that spanned at least six lines of data were preferred, thus reducing the six-line striping effect (Chapter 6). Mixtures of grass, grass and clover, or grass and bare soil were sought. The panchromatic aerial photography was used to check their uniformity (table 14.2).

(iii)   Histograms were prepared of the radiance level occurrence statistics for each MSS channel and class. Only those pixels located at least one pixel width away from the field boundaries were included, thus avoiding edge effects. The upper and lower limits of radiance for each MSS channel and class were obtained by examining these histograms.

(iv)   The mean values in each MSS band and class were derived. The values were plotted at a convenient scale to show interchannel trends within each class. This was done for each class and for different subclasses due to crop condition, date of planting, amount of grazing etc. As a result a series of mean curves was produced that reflected the variations within the class due to the above factors.

(v) Spectral signatures were defined for each Level III class by selecting the upper and lower limits in each MSS band. These were chosen to either represent that class or to break the class into a series of spectrally overlapping subclasses. This latter course was sometimes necessary to reduce overlap with other classes or subclasses. In each case, the radiance level 'gates' were widened by one or two radiance levels at either end, to take noise due to striping into account. The spectral signatures resulting from this analysis are presented in figure 14.2.

(vi) A parallelepiped classification module (Thomas 1979) was then entered with the above spectral signatures and operated as described in Chapter 6. Spectral signatures of subclasses at Levels IV and V (table 3.2) were allowed to overlap, since the objective was to aggregate these into a Level III classification.

(vii) Output products such as the thematic map of plate 6 and character-coded classification maps (as, for example, figure 12.3) were obtained.

Table 14.5 Darfield study area—classes used in parallelepiped classification.

1		Type *A*	All classes include grass, clover, lucerne, and
2	Pasture	Type *B*	mixtures of grass and clover, and wheat stubble
3		Type *C*	and clover. The distinction among the classes is
4		Type *D*	that from *A* to *D*, paddocks go from well-maintained, vigorous pasture to either run-out or newly seeded pasture with a high proportion of bare ground showing.
5		Type *A*	These are predominantly wheat. The distinction
6	Cereal crop	Type *B*	among these classes is that from *A* to *D*,
7		Type *C*	paddocks go from vigorous, early-sown spring
8		Type *D*	wheat to late-sown spring wheat (and barley) with a high proportion of bare ground showing. Autumn wheat and oats occur largely in class *C*, along with the spring wheat.
9	Rape		These are recently worked paddocks, either
10		Type *A*	completely bare or with crops no more than a few
11	Bare ground	Type *B*	inches high. Potatoes and peas generally fall into
12		Type *C*	types *A* and *B*, while spring wheat and barley generally fall into types *B* and *C*. Run-out pasture with very high proportions of bare soil may also be caught (rarely)in these classes.
13	Forest		
14	Water		

(viii)   The classification results were checked against ground truth data collected from different areas to those used in actually setting up the spectral signatures.

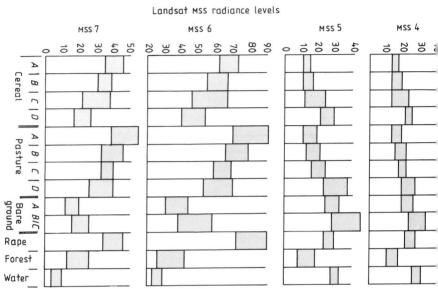

Figure 14.2 Spectral signatures used in the parallelepiped classification.

14.5.2 Results

Using the October scene, the Darfield region was classified between pixels 817–1632 and lines 1615–2202. Most of the pixels in the study area were classified into the above classes using the 'gates' of figure 14.2. Pixels still remaining unclassified were nearly always other types of bare ground, for example braided river gravels.

Areas of scrub, river meadows, the town of Darfield, and other uncultivated areas were also classified as one of the 13 classes in table 14.6. Therefore any area measurements using the total number of pixels in each class for the whole study area were invalid, being much too high as a direct consequence.

Although ground truth was originally obtained for approximately 200 fields, the results presented here have been reduced to 74 fields. These were the ones which were common to the subsequent Maximum Likelihood analysis (§ 14.6). This enabled a direct comparison to be made between the parallelepiped classifier and Maximum Likelihood results. In table 14.6 the field classifications are based on a majority decision, with each field being assigned to the class represented by the majority of pixels within its boundaries.

The results in table 14.6 indicate that misclassification can exist between autumn and spring wheat, between all pasture types and between spring barley and bare ground. Other crops such as peas did not have an individual class and were also classified in the bare ground group. During the month of data acquisition (October), spring barley had a high proportion of bare ground to plant cover, and therefore had a spectral signature similar to that of bare ground.

At Level III (table 3.2) (i.e. wheat, pasture and bare ground), the classification is in error in only five out of a total of 74 fields—utilising the majority decision approach.

In some cases, an overlap exists between two classes in all four MSS bands (figure 14.2). For instance, six sets of radiance levels are common to cereal *B* and cereal *C*. In this situation, the parallelepiped classifier will assign pixels with these signatures in an arbitrary fashion, using an order of precedence based upon the order in which the 'training' signatures are inserted into the system.

Because of the small number of fields involved in this 'fields only' classification, it is not possible to derive a confidence level for the accuracy of the results (Chapter 11). However, the results do serve to indicate the *level* of classification that might be achieved, i.e. Level III (table 3.2).

Table 14.6 Numbers of fields classified by parallelepiped classifier using the majority decision criteria.

Ground truth classes	Spectral classes												
	Cereal				Pasture				Bare ground				Rape
	A	B	C	D	A	B	C	D	A	B	C	D	
Autumn wheat		5	8										
Spring wheat	1	3	3										
Wheat group	1	8	11										
Lucerne					9								
White clover					3	1							
Growing pasture					10								
Grazed pasture					9								
Pasture group					31	1							
Fallow (ploughed)			1		1						2	4	
Peas												6	
Spring barley		2	1							1	1	2	
Bare ground group		2	2		1					1	3	12	
Rape seed													2

14.6 The Maximum Likelihood Classification

14.6.1 Method

The Maximum Likelihood approach used in ERMAN and the analysis pathway used by the New Zealand group is outlined in Chapters 7, 8 and 9. Additional aspects of particular concern to this agricultural study are presented below.

14.6.1.1 Registration

Starting with the Landsat computer tapes of the August and October scenes (table 14.4), a subscene was chosen to cover the Darfield study area. In order to merge the Darfield subscenes for August and October, they were registered with 40 m resampling to obtain a pixel to pixel correspondence. (Ground Control Points were derived from the NZMS 1 topographic map at 1:63 360 scale (see Chapter 8).) The resampled images have a pixel size of 40×40 m², compared with the original Landsat pixel size of 79×56 m². Resampling to a smaller pixel size is desirable when rectifying an image in order to preserve the shape of small targets. This is particularly necessary if the registration process involves an image rotation.

14.6.1.2 Selection of training fields

Training fields are used to compute statistics to characterise each class. For the Maximum Likelihood classifier these are the mean radiance values for each MSS band and the variance/covariance matrices (see Chapter 7).

These fields were most easily located using the colour display screen and an enhanced colour composite of MSS bands 4, 5 and 7.

The criteria for choosing a particular field were the available ground truth together with the apparent homogeneity of the field as displayed on the colour monitor. A wide variation of colour within a field corresponds to a large spread of radiance levels in one or more of the data channels. Homogeneity is considered to be desirable since the class is then closely defined and the likelihood of overlap with other classes is reduced.

The training fields are intended to be representative samples of the entire class. If the class itself is not homogeneous, then the training fields should reflect this inhomogeneity. This may be seen in the shape of the class boundary in feature space being irregular, and thus not fitting easily into the boundary set by a particular threshold level of the Maximum Likelihood classifier. This is equivalent to saying that, at least in some dimensions of feature space, the distribution of points within a major class may not follow a Gaussian distribution.

In this situation, our practice was to split the class into two or more spectral subclasses and then to recombine these classes when the classification was concluded. Such subclasses were chosen at a level where homogeneity could be supported and thus the parametric classifier approach was followed (see Chapter 7).

14.6.1.3 Use of Synthetic Channels

The generation and use of Synthetic Channels is discussed in detail in Chapter 6. For these agricultural investigations, all the Band Ratios and Principal Component channels were generated and added to the basic data set for subsequent channel selection using the Divergence option (see Chapter 7).

14.6.1.4 Classification

Three classifications of the Darfield test area were made. The first used an image set consisting of the original four MSS bands of the October scene. The second classification used fifteen channels selected from the full set of MSS plus Synthetic Channels for the October image. The Divergence option (Chapter 7) was used to make the selection. The third classification utilised the four MSS bands from the August scene, and the four bands from the October scene, giving an eight-channel classification based on time-separated imagery. These three classifications of the same scene permitted a comparison to be made of the accuracies of the three methods of analysis.

For the second classification (above) of the fifteen channels: Divergence was used to find the best eight out of the set of twenty channels over all classes. The average Divergence between pairs of classes was then examined using eight channels, and the weakest (least separable) pairs of classes were identified. Those channels giving the best separability between these weak pairs were then added to the set of four, making up the total of fifteen channels. Table 14.7 lists the channels selected.

Table 14.7 MSS and synthetic channels selected by the Divergence option (PC = Principal Component). For details of the ratio algorithm, see Chapter 6.

Channel number	MSS bands	Selected by divergence	Channel number	MSS bands	Selected by divergence
1	4	★	11	'6/5'	★
2	'4/5'		12	'6/7'	★
3	'4/6'	★	13	7	★
4	'4/7'	★	14	'7/4'	★
5	5	★	15	'7/5'	★
6	'5/4'	★	16	'7/6'	
7	'5/6'	★	17	PC1	
8	'5/7'	★	18	PC2	★
9	6		19	PC3	
10	'6/4'	★	20	PC4	★

A total of 29 classes were used in these classifications, of which only nine were used in the agricultural analysis. The remainder were general land cover types such as water and river gravel, and a set of forest classes.

14.6.1.5 Output products

ERMAN provided a variety of output products (Chapter 10). For these investigations we obtained the following:

(i) A character-coded map, produced by the laser lineprinter, showing the four MSS band classification. For one output run no lower limit (threshold) was set to the likelihood value, and therefore all the pixels were included into one or other of the specified classes. (See Chapter 7 for comments on thresholding.)

(ii) A character-coded lineprinter product of the same classified data but with a 1% threshold level. This eliminated all pixels with a less than 1% likelihood of belonging to any of the nominated classes.

(iii) A character-coded map of the fifteen-channel classification with no threshold.

(iv) The same as in (iii) but with a 1% threshold.

(v) A character-coded map of the eight-band August and October combination with no threshold.

The classification results were also copied to magnetic tape with all classes given a coded value, which were subsequently used to produce colour-coded thematic maps on the Colorwrite machine (see Chapter 10).

In addition to the above products a magnetic tape containing the classification dataset was obtained. Copies of the statistical data associated with the project were also prepared to assist in other supporting analyses.

14.6.1.6 Classification results

In addition to compiling a set of training fields from the available ground truth, a further set of test fields was required to estimate classification accuracy. Such a test field set should always be at least as large as the training field set. It is not statistically acceptable to include training fields within the test field set. Similar ground truth data are required for the test fields as are used for the training fields.

The original set of 200 fields for which ground truth was obtained yielded a sub-set of 99 fields for training and subsequently checking the classification results. Reasons for the reduced figure included the difficulty of boundary determination for some fields with irregular boundaries, unreliable ground truth and the fact that some outlying fields were outside the area covered by the detailed study. The total of 99 fields was divided into 24 training fields and 75 test fields.

The number of pixels correctly classified in each class is given in table 14.8, for each of the two classifications using the October Landsat database. This table gives the results when a 1% threshold is used to exclude those pixels with a less than 1% likelihood of belonging to a nominated class.

Table 14.8 Number of pixels correctly classified from the October Landsat data in each class and group. Class boundaries are set by the decision boundaries or the 1% threshold, whichever is the greater. NB. The 'group' results permit misclassification between subclasses of that group.

Class and group	Total pixels	Number of pixels correctly classified	
		4-channel	15-channel
All classes	4520	1053	403
All groups	4474	2647	1276
Autumn wheat	746	334	140
Spring wheat	328	13	0
Wheat group	1074	591	245
Lucerne	473	61	55
White clover	369	68	45
Growing pasture	772	81	57
Grazed pasture	685	194	17
Pasture group	2299	1543	808
Fallow (ploughed)	372	160	27
Peas	321	60	3
Spring barley	408	82	59
Bare ground crop	1101	513	223
Rape seed	46	0	0

Table 14.9 gives the results of a majority decision classification of fields, derived from the data when a 1% threshold is used. In the majority decision approach it was assumed that each field boundary was accurately known, and that the field was allocated to the majority class within the boundary.

Two points emerge from assessing the data in tables 14.8 and 14.9. The first underlines the hierarchical nature of classifications. The classification by groups has a higher percentage correct score than the classification into the subclasses of that group.

The second point is how the selected training fields actually represent the variability in the nominated classes over the classified area. A comparison between the four-channel and the fifteen-channel classifications indicates a decline in the percentage correct scores as we go to the fifteen-channel classification. The four-channel likelihood distribution is usually less tightly confined (i.e. it has a greater standard deviation) for each class than the fifteen-channel distribution. As a result, more errant subclasses are excluded from the likelihood region that belongs to the finally accepted ('thresholded in ') fifteen-channel classification than from the similarly thresholded four-channel classification. A more exhaustive search for representative training fields for each subclass coupled with the amalgamation of their statistics into a broader dataset for that composite subclass should make the resultant classification more representative over a wider area.

Table 14.9 Field classification by majority decision for the two October Landsat data classifications invoking a 1% threshold boundary. NB. The 'group' results permit misclassification between subclasses of that group.

Class and group	Total fields	Number and % of pixels correctly classified			
		4-channel		15-channel	
All classes	75	27	36 %	25	33 %
All groups	74	68	92 %	60	81 %
Autumn wheat	12	7	58 %	9	75 %
Spring wheat	5	0	0	0	0
Wheat group	17	13	76 %	14	82 %
Lucerne	8	2	25 %	3	38 %
White clover	8	2	25 %	2	25 %
Growing pasture	12	1	8 %	1	8 %
Grazed pasture	9	3	33 %	2	22 %
Pasture group	37	37	100 %	33	89 %
Fallow (ploughed)	8	5	63 %	5	63 %
Peas	6	3	50 %	1	17 %
Spring barley	6	4	67 %	2	33 %
Bare ground group	20	18	90 %	13	65 %
Rape seed	1	0	0	0	0

If we now accept that we have some very small fields in this agricultural environment compared with the original sampling pixel size of Landsat, we can consider replacing the (at least) 1% threshold requirement with the inclusion of field boundary qualifying information. In our so-revised assessments we confine ourselves to all those pixels within the boundaries of the fields being used to check the classifications. That is all pixels—no imposition of a threshold—are evaluated for being either correctly or incorrectly classified within the evaluatory dataset. This leads to the pixel data presented for the three classifications in table 14.10.

Table 14.10 Number of pixels correctly classified in each class and group for each of the three classifications. Class boundaries are purely decision boundaries, i.e. no likelihood value threshold has been set.

Class and group	Total pixels	Number of pixels correctly classified		
		4-channel	15-channel	8-channel
All classes	4520	1351	1368	1400
All groups	4474	3229	3253	3428
Autumn wheat	746	403	616	342
Spring wheat	328	26	1	33
Wheat group	1074	751	927	647
Lucerne	473	67	122	78
White clover	369	92	132	190
Growing pasture	772	99	87	131
Grazed pasture	685	258	83	142
Pasture group	2299	1803	1720	1891
Fallow (ploughed)	372	176	139	169
Peas	321	80	22	40
Spring barley	408	150	166	275
Bare ground group	1101	675	606	890
Rape seed	46	0	0	0

By now applying the confidence level concept of Chapter 11 and therefore allowing for sample size the comparative confidence estimates for the three classifications can be deduced and are presented in table 14.11.

Table 14.11 99.9% percentage confidence levels for classification by pixel at 0% threshold.

Class and group	Total No pixels	Number of pixels correctly classified (%)		
		4-channel	15-channel	8-channel
All classes	4520	27.3	27.6	28.3
All groups	4474	69.6	76.3	74.1
Autumn wheat	746	47.4	77.4	39.2
Spring wheat	328	2.2	0	3.7
Wheat group	1074	64.8	82.4	54.8
Lucerne	473	8.2	18.4	10.1
White clover	369	16.6	26.6	42.0
Growing pasture	772	8.3	7.0	12.0
Grazed pasture	685	31.0	7.4	15.0
Pasture group	2299	75.2	71.4	79.2
Fallow (ploughed)	372	37.8	28.1	35.9
Peas	321	15.9	1.4	5.5
Spring barley	408	28.0	31.8	58.9
Bare ground group	1101	56.0	49.6	76.4

14.6.1.7 Discussion on Maximum Likelihood classification results

A number of conclusions may be drawn from the foregoing results:

(i) The Level III agricultural ground cover has spectral signatures in Landsat (1, 2, 3) data which fall into the natural groupings of wheat, pasture and bare ground.

(ii) Wheat can be separated from the background using a single (spring) image only, with a 99.9% confidence limit (Chapter 11) of 82% for the fifteen-channel classification (of pixels), and 65% for the four-channel classification.

(iii) If a majority decision approach is used to classify fields in which the boundaries have been previously determined, the confidence level approach cannot usually be used because of the small number of samples in each class. An advantage of the technique is that all areas which are not fields do not enter into the classification, and cannot therefore contribute spurious pixels into the agricultural classes.

(iv) Combination of the winter and spring images in a single classification did not significantly improve the results. An improvement was noted in pasture, which probably did not change greatly between winter and spring. For wheat, the classification accuracy appears to degrade. This may be due to the presence of other crops in the same field at the time the winter image was acquired (see table 14.1).

(v) Cereals such as barley were easily separated from wheat at the time of the spring image. This appears to be due to the high proportion of bare ground within the barley field at that stage of its growth, resulting in classification of barley fields into the bare ground group. In summer the barley spectral signature would be different, and perhaps more like wheat.

(vi) In both the winter and spring images the spectral signature differences between the pasture types are not very great. In late summer grass pastures dry out, white clover cash crops are harvested and only lucerne (alfalfa) tends to retain its vigour because of its deep rooting system. It is believed that a late- or mid-summer image would enable these classes to be differentiated much more clearly.

(vii) Often one of the objectives of the classification process is to measure the total area of a given crop within a region. A total pixel count of that class does not give an accurate result for the following reasons:

(a) The small field size (on average 9.6 ha) means that a high proportion of Landsat 1, 2, 3 pixels within a field are influenced by the boundary. In the case of the Darfield test area the average proportion of boundary pixels within a field is 46% (table 14.3). These pixels can be misclassified because of the influence of their surroundings on the spectral signature.

(b) The results of tables 14.8 and 14.10 show that the introduction of even a 1% threshold greatly reduces classification accuracy within known fields, because of the exclusion of many pixels. However, if no threshold is used, then all pixels in the image must be allocated into one or other of the specified classes in the output products. This usually results in an overestimation of the numbers of pixels in a specific class. For instance, pixels lying on a boundary between forest and pasture have been classified as autumn wheat. If there is no specific class for urban areas, these will tend to be classified as bare ground. Unless the threshold levels can be carefully adjusted to the situation, errors of omission will occur if the threshold is too high, and errors of commission will occur if no threshold is used.

It is for these reasons that the majority decision approach is suggested as potentially the most accurate remote sensing method of measuring crop areas within a region. A map of field boundaries and an accurate estimate of each field area is a necessary requirement before the satellite image can be analysed. It is necessary to register the map and image before the majority vote for each field can be obtained. The result is a list of fields of known area, each assigned to a member of the set of classes.

An advantage of this process is that the classification can be made with no threshold, thus avoiding errors of omission. Areas which are not previously identified as fields are automatically excluded from the analysis, thus avoiding errors of commission, caused, for instance, to wrongly classifying forest–pasture boundaries as wheat. Moreover, the degrading effect of boundary pixels is much reduced, since the majority decision will favour the correct classification. For the previously determined field set, the area estimate for a given crop must be more accurate than by pixel counting methods, since boundary pixels will not influence the result.

(viii) These investigations have shown the need for accurate location of targets on a Landsat image. This is particularly true in a mixed farming area with small fields like Darfield. It is believed that with future high resolution satellite imagery (10×10 m^2), field boundaries and areas may be deduced directly from the satellite image. This will have the additional benefit of enabling maps of the area to be updated very regularly. The majority decision approach can still be used to eliminate errors of commission, although this implies that the field boundary data set must also be updated regularly.

(ix) For this project, the set of training pixels used to generate the statistics for each class was derived by choosing a small number of training fields for each class. Consequently the statistics are compiled from pixels selected by drawing boundaries around each training field on the display screen. The results of these investigations show that this method of compiling a set of training pixels does not necessarily produce a set which is truly representative of the desired class. A better method would be to select samples of pixels for the training set from *within each field* for which ground truth is available. These sample pixels would then be excluded from the testing process. This approach would generate a set of classes with broader standard deviations, but a better chance of being more representative of their class.

An equivalent approach would be to take samples, constructed by bounding the sampled pixels with a user defined boundary, over a wider range of variants in each of the subclasses. These variants would cover the geographic area to be classified, the range of soil types/microclimates to be embraced as well as the variety of farming practices etc. Obviously the bounding contour would again be set into the field away from the edge effects.

To an extent, this approach has been followed here with the sampling being conducted at a Level V or VI detail (see Chapter 3). Level III can seemingly be attained but most of the problems in species differentiation occur at Level IV.

14.7 Comparison of Parallelepiped and Maximum Likelihood Classifications

The two sets of results do show that the spectral signatures for the two approaches fall into the same groupings. The implication is that the parallelepiped classifier can achieve a Level III classification from imagery taken in spring. Thus it is comparable to the Maximum Likelihood classifier if the only objective is to classify targets at this 'macro' level. At more refined levels, the superior characteristics of the Maximum Likelihood classifier give better results. For the parallelepiped classifier, a higher degree of human input and expertise is required in determining the radiance level limits in each band. This factor can reduce the reproducibility of the results.

14.8 Summary and Conclusions

One of the earliest objectives of the Darfield investigations was to discriminate major crops, such as wheat/barley, from one another. Within the test area at least, this has been shown to be feasible, even with only one image taken in spring.

Another objective was to determine the classification level at which targets can be reliably separated. For images taken in winter and spring, it is seen that only the major groups can be routinely separated using either the parallelepiped or Maximum Likelihood classifier (Level III classification). It is believed that some Level IV classes could also be separated, such as the pasture classes, if summer imagery were available.

For the major groups, a high classification accuracy has been achieved, when combining the majority decision (classification by field) approach with the conclusions advanced by the Maximum Likelihood analysis.

14.9 References

Anderson J R, Hardy E E and Roach J T 1972 *A Land-Use Classification System for Use with Remote Sensor Data* USGS Circular No 671 (Washington, DC: US Geological Survey)

Ellis P J, Thomas I L and McDonnell M J 1978 (ed.) *Landsat II Over New Zealand* Bulletin No 221 (New Zealand: Department of Scientific and Industrial Research) ch 2

New Zealand Official Yearbook 1981 (Wellington, NZ: Department of Statistics) Section 14

Thomas I L 1979 Cartography from Landsat: introducing the DSIR computerised land use mapping package *Proc. 10th New Zealand Geography Conf. and 49th ANZAAS Congr.—Geographical Sciences (Auckland, NZ)* p 276

15 | Classification of Agricultural Land Cover from Aircraft Scanner Data

15.1 Introduction

During the 1980/81 austral summer, an aircraft scanner survey was executed over an area around Darfield (see figure 14.1). The aim was to map the different types of agricultural cover using the aircraft scanner data and compare the results with those of the previous Landsat agricultural study (Chapter 14). It was hoped that the improved spectral and spatial resolution of the aircraft scanner data would produce a more detailed classification than that derived from Landsat data.

A second aim was to indicate the type of information the next generation of satellites might provide. The farming industry is important to New Zealand's economy and particularly to export earnings. Thus, if future satellite data could contribute to farming efficiency the benefits could be significant. The multispectral data were collected at a nadir pixel size of 10 m, i.e. with an overall resolution of between 10 and 20 m. Further details on the aircraft scanner system used are presented in Chapter 5.

The flight was made on 9 January 1981 at 09.30 NZ Summer Time. The actual aircraft scanner project test area is portrayed by the images in plate 7. The images are various three-band combinations of the Darfield aircraft scanner data, as explained in the caption.

15.2 Method

15.2.1 Ground truth

Farms within the Darfield area were visited in late November 1980, and also on the day of the scanner overpass. The following ground truth information was collected:

(i) crop type and variety and stage of maturity, moisture availability;
(ii) past fertiliser or other chemical treatments;
(iii) the presence of disease or weeds.

Having collected the data, a hierarchical format for information recording was devised (figure 15.1). Thus, each paddock was described by

232

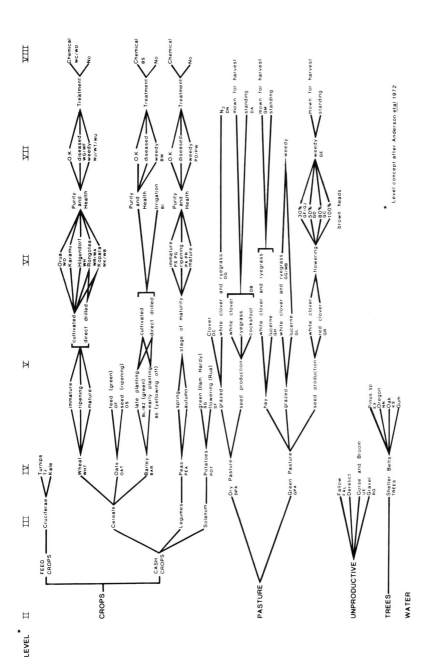

Figure 15.1 Hierarchical format for ordering the agricultural information appropriate to the aircraft scanner study.

a *set* of names, for example:

Cereal—wheat—ripening—rongatea—wild oat infestation
Pasture—green—grazed—white clover/ryegrass.

15.2.2 Selection of training and test fields

At least one training field was required for every cover type to be mapped from these data. The possible number of classes at Level VII (figure 15.1), the initial classification level attempted, was rather large and the ERMAN software package, the analysis system used here, could only accept 60 discrete classes. Therefore, crop types which occurred in only one or two paddocks were culled from the potential class list. One to three training fields were finally chosen for each of 59 classes (those marked with abbreviated names on figure 15.1).

Between one and nine non-training fields per map class were also selected for later use in testing the accuracy of the classification. Where possible, the selection was made so that the test fields for a class were distributed throughout the image, i.e. under different sun/scanner angle conditions, and possibly different soil influences. Only fields which were visited on the day of scanner data collection were chosen as training or testing fields in order to minimise the impact of farm practice variations on the basic crop description.

15.2.3 Aircraft scanner data corrections

Unacceptably bad radiance level shifts (banding) were present in the Darfield aircraft scanner data. These were caused by failure of the gain correction system of the scanner during the actual scanner flight. Consequently they necessitated the correction of the raw data before they could be used for digital processing. The problem was largely alleviated by using new offset and gain corrections. These were determined from the calibration data at a point in the survey where the scanner was known to be operating correctly. Details of this correction are given in Chapter 5.

The uncorrected Darfield data appeared to have two horizons. This 'S-bend' distortion, characteristic of an instrument scanning a relatively wide scan angle with a fixed Instantaneous Field of View, was also corrected, as described in Chapter 5. The result was an image whose pixels across the scanline were a constant size, with reference to the ground.

It was these spectrally and spatially corrected data that were used in classification.

15.2.4 Classification of the aircraft scanner data

The ERMAN system was used to classify the aircraft scanner data. Its operation is described in Chapter 9, as are the details of the analysis pathway followed here.

15.2.4.1 *Entry of training fields*

The first step was to enter the 77 previously selected training fields into the ERMAN system via the colour display screen.

The training set was checked by running a classification accuracy test on the training fields. Using a 0% threshold, an average of 99.2% of the pixels in the training fields for all classes were classified correctly in the set of training fields finally selected.

15.2.4.2 *Channel selection*

The eleven aircraft scanner channels (see Chapter 5) were used in the classification of training fields. It seemed unnecessary to create further synthetic channels when eleven independently collected bands were available. The Divergence option run for the Darfield agricultural classes showed that class separability improved as the number of channels increased (Chapter 7). However, this trend reached a plateau at about eight channels for this agricultural project suggesting that six to eight channels may well have been the most cost-effective number of channels to use (see figure 7.8). However, it was felt desirable to retain *all* the original data channels and thus the full aircraft scanner channel set was used in the classification.

15.2.4.3 *Inclusion of evaluatory test fields explicity in the classification*

The other set of fields for which reliable ground truth existed was also entered into the ERMAN system in the same manner as the training fields. Once the whole scene was classified using the eleven data channels, the individual test fields were then classified using the same statistics. The number of pixels within each test field assigned to each class was then reported. The whole map classification accuracy was deduced from the statistics of the 111 test fields for both the 0% and 1% threshold maps produced.

This method of assessment of classification accuracy was found to be far more efficient than that described in Chapter 14 since the pixel counting was done by the computer, rather than the human analyst.

15.3 The Classification Map and its Accuracy

15.3.1 *Test field results*

Table 15.1 presents the accuracy figures for the classification with a 0% threshold applied, derived directly from the classified test field results. The percentage of pixels in each test field assigned to each of the 59 possible classes is given.

Table 15.1 Darfield aircraft scanner classification accuracy results for Level VII, 0% threshold map. Percentage of test field pixels assigned to given classes. The symbols are as given in figure 15.1.

Test fields	No test fields	No test pixels	Symbol	Potatoes		Oats		Wheat										Barley						Peas	
				SG	SF	OF	OS	WR	WH	WK	WA	WB	WC	WJ	WF	WT	WU	BZ	BE	BI	BS	BW	BM	PS	PO
Potatoes	1	296	SF	15	0			5			2								3							18
Oats	1	685	OS				98																		
Wheat	8	6185	WR			2		46	5	8	8	3	10	8						2					
	2	2399	WH					6		8				52		4	23			1					
	7	6334	WK						15	8															
	3	3117	WC					12		46		23	5	5	1	1	5		1	1					
	3	1024	WF					3			34				14					43					
	3	4673	WT					15			3	15	4	13		7	22			1					
Barley	4	2609	BZ			2												1	37	21	12	18	35		
	8	7786	BE				1												17	12	12	15			
	4	3359	BL			8		6										17	2	17	7				
	1	544	BS																				99		
Peas	4	3474	PS																					9	
	1	1278	PW								2									7					9
	3	2148	PG																						
Dry pasture	7	6943	DG																						
	11	9367	DS					5				7													
Green pasture	9	9032	GG																						
	6	5325	GL																						
	2	1958	GH			15		1											2	11					
	3	2047	MG																						
	1	1049	GD																						
	3	2128	GE			3																			
	3	2849	GF																						14
	3	1322	GR		2															1					
Fallow	10	7251	FL																						

Table 15.1 (contd)

Column codes: PW, PA, PG, PR | Dry pasture: DS, DN | DE | Green pasture: GG, GW, GH, GC, GE, GF, GJ | MG, IR, NR | FL = Fallow, VL = Gravel, UE = Gorse, T = Trees

Test fields	No test fields	No test pixels		PW	PA	PG	PR	DS	DN	DE	GG	GW	GH	GC	GE	GF	GJ	MG	IR	NR	FL	VL	UE	T
Potatoes	1	296	SF			27										3								21
Oats	1	685	OS			1							1											
Wheat	8	6185	WR									1												
	2	2399	WH																				3	6
	7	6334	WK																					
	3	3117	WC											1									2	
	3	1024	WF																					
	3	4673	WT								1													
Barley	4	2609	BZ			6						2												2
	8	7786	BE								12													17
	4	3359	BL													2								24
	1	544	BS																					
Peas	4	3474	PS	36		34						7												
	1	1278	PW	1	64		87																	
Dry pasture	3	2148	PG		1	2					9	3	1			11		6						4
	7	6943	DG	1				4	67		2					1		4						
	11	9367	DS	1	1	2		45	23	2								2	2	3				
Green pasture	9	9032	GG						13	18	28	6	18	11		27								
	6	5325	GL								16		34			7		30						3
	2	1958	GH								63		32											
	3	2047	MG								19	9	6					15						
	1	1049	GD								17		16				42			4				
	3	2128	GE			3						32	12		18	3		3						
	3	2849	GF										14			32								
	3	1322	GR			7					3	2	29			36	12		4					
Fallow	10	7251	FL																		98			

The use of a 0% threshold here for the discrete test fields is considered to be equivalent to the 'majority decision' approach employed in Chapter 14 for well-defined fields.

Clearly the classification accuracy at Level VII (figure 15.1) is poor. The only exception is the oats class, for which only one test field was provided. However, although there are a lot of misclassifications, most are intra- rather than inter-group mistakes, shown by the strong lead diagonal in table 15.1.

As the trend of table 15.1 would suggest, amalgamation of the classes to Level V (figure 15.1) improved the accuracy of classification of most classes (table 15.2). At this level, oats, late planted barley, autumn peas, grazed dry pasture and fallow classes were satisfactorily classified. However, a lot of early barley pixels were classified as late barley, spring peas as autumn peas, and pasture saved for seed as grazed dry pasture.

At Level V therefore there was still too much confusion between classes, particularly allied classes. The number of pixels correctly classified was often matched by incorrect classifications. Consequently the classes were aggregated further.

The accuracy figures for the Level IV classification are given in table 15.3. At this level, the mapping accuracy is very good for all classes. The overall probability of correct classification at Level IV was 79% with 99.9% confidence limits from the 0% thresholded test field data.

The Level IV classification is given in plate 8. It was written out on a Colorwrite machine translating *class* number (1–59) to the appropriate *group* colour (eleven colours) via an appropriate look-up table.

15.3.2 Thresholding option

In the 0% threshold classification every pixel was classified into one of the specified classes, whether or not a truly appropriate class was provided. Hence the urban areas, such as Darfield township, were mapped as flowering potatoes because no urban class was included in the training set. Using a 1% threshold however, pixels are only included in a class if they have a likelihood of inclusion greater than 1% of the peak likelihood value.

According to statistical theory, a threshold should always be applied when a Maximum Likelihood classifier is used. This is because a 0% threshold implies that feature space is completely filled and that a training field has been provided for *every* ground cover type present. With eleven-dimensional data and only 59 training classes used here obviously all possible spectral and ground classes could not be accommodated (see Chapter 7).

Although almost all confusion between classes disappears using the 1% threshold, so does most of the classification. Therefore the 0% threshold map given in plate 8 was selected as the final map, despite the theoretical difficulties.

Table 15.2 Darfield aircraft scanner classification accuracy results for level V, 0% threshold map. Percentage (%) of test field pixels assigned to given classes.

Classification

Test fields	No test fields	No test pixels	Potatoes	Oats	Wheat ripening	Wheat mature	Wheat wild oats	Barley early	Barley late	Peas spring	Peas autumn	Dry pasture grazed	Dry pasture seed	Green pasture grazed	Green pasture hay	Green pasture seed	Fallow	Gravel	Gorse	Trees
Potatoes	1	296	15	7	5	2			3	18	27					3				21
Oats	1	685		98							1									2
Wheat ripening	17	14918		1	33	7	29		1										1	
mature	3	3117			58	24	11		1										2	
wild oats	6	5697			32	22	42		1										2	2
Barley early	8	7786						17	65					12						
late	9	6512						13	64		2									14
Peas spring	5	4752		3	2	1		2	2	23	44		1	3	1	1				
autumn	3	2148		2	1				2	10	65			14	4					4
Dry pasture grazed	7	6943			5	7			7	1	1	74	3	6	1					
seed	11	9367									3	68	3	1		11				
Green pasture grazed	18	16904						2			3	4	2	46	20	19				
hay	2	1958		15	1				1	1			18	17	32					
seed	10	7348	1	1					11					20	18	49				1
Fallow	10	7251	1		1					4		1					98			

Table 15.3 Darfield aircraft scanner classification accuracy results for Level IV, 0% threshold map. Percentage of test field pixels assigned to given classes.

Test fields	No test fields	No test pixels	Classification								
			Potatoes	Oats	Wheat	Barley	Peas	Dry pasture	Green pasture	Fallow	Trees
Potatoes	1	296	15	7	7	4	45	3			21
Oats	1	685		98			1		1		1
Wheat	26	24230		3	75	8	2				10
Barley	17	14298			2	78		1	4		2
Peas	8	6900			1	4	80		10		
Dry pasture	18	16310			6		2	74	12		
Green pasture	30	26210		2	1	2	3	5	84		
Fallow	10	7251								98	

To avoid the problems described above and to enable the thresholding option to be used effectively the combination of spectral signatures at Level VII to Level IV should have been employed. This would have resulted in an aggregated class—as we've just outlined and demonstrated in tables 15.1, 15.2, 15.3. As such the aggregated class statistics would have better represented the breadth of Level V, VI, VII influences upon the Level IV basic classes. Time acted against this extension in this study. If this technique were used the thresholding could have been validly employed and would have led to a more representative classification product.

So, from an operational standpoint, it is considered advisable to run the classification at Level VII, or whatever level is selected; then consider the amalgamation to Level IV (in this case) but then aggregate the class statistics to that level and repeat the classification.

15.4 Conclusions

The results of this study clearly show that aircraft scanner data recorded during the summer season just prior to harvest (see Chapter 14) can be used to discriminate the cereal crops: wheat, barley and oats. Furthermore, the data can separate dry from green pasture, and can map bare or fallow areas exceedingly well. Although discrete spectral signatures were obtained for Level VII classes, the data did not support classification of the whole image at this level.

Overall pixel classification accuracy to the 99.9% confidence limits (Chapter 11) from the 0% thresholded classification dataset was 79% at the Level IV class amalgamation.

15.5 References

Anderson J R, Hardy E E and Roach J T 1972 *A Land-Use Classification System for Use with Remote Sensor Data* USGS Circular No 671 (Washington, DC: US Geological Survey)

Glossary

Reference should be made to the index for terms not included in this glossary.

Other useful references are the glossaries in the *Manual of Remote Sensing* (2nd edn) (American Society of Photogrammetry, 1983), *Remote Sensing: The Quantitative Approach* edited by Swain and Davis (McGraw-Hill, New York), and the *Dictionary of Computing* (Oxford Science Publications, Oxford, UK, 1984).

Absorption: the process by which a specific material absorbs radiant energy (such as light) and converts it into another form of energy.

Accuracy estimate: a measure of the similarity of an estimate to the true value.

Advanced Very High Resolution Radiometer: AVHRR, a 4/5 channel imaging sensor package flown on the National Oceanic and Atmospheric Administration (NOAA) satellites. It measures the intensity of electro-magnetic radiation in selected wavelength bands.

Aerial photography: photographs which are taken from an aircraft. These photographs are generally taken vertically towards the ground and so that adjacent photographs have an overlap. (This overlap commonly ranges from 20 to 60%.) The overlap along the flight path is used to obtain a stereoscopic view for interpretation or mapping purposes.

Alphanumeric: character set consisting of letters and numbers. Alpha-numeric lineprinter: a terminal connected to a computer which prints, on paper, lines of letters and numbers.

Altitude: height above a reference level.

Artificial Intelligence: AI, taken as that approach that moves a computer system towards performing those tasks that when done by a human require not only intelligence but perception too. For example, matrix inversion is not considered a candidate for AI whereas speech and vision recognition systems are. The key element in the perceptual areas is knowledge that has been imparted to the system (human or computer) through training.

Atmospheric extinction coefficient: atmospheric absorption is usually wavelength dependent and the amount of the absorption can be assessed for various wavelength bands. Such coefficients are known as atmospheric extinction coefficients. The concept is also used frequently in astronomy.

Band: (i) a range of wavelengths; (ii) a group of wavelengths in the electromagnetic spectrum; (iii) a series of intensity values measured and stored on magnetic tape within a selected wavelength interval.

Band Ratio: this is a data channel which has been created by dividing the multispectral value of one band into the multispectral value of another band for each pixel in a scene. The algorithm that produces the Band Ratio usually includes a factor to avoid resulting negative values. These bands are often referred to as synthetic bands.

Bayes decision rule: a statistically based algorithm for classifying individual pixels into their appropriate classes.

Bit: Binary digit. A single character (0 or 1) of the computer language.

Bottom-up coding: this is an approach to program development in which progress is made by composition of available elements, beginning with the primitive elements provided by the implementation language and ending when the desired program is reached. At each stage the available elements are employed in the construction of new elements that are more powerful in the context of the required program. These new elements will in turn be employed at the next stage in the construction of still more powerful elements and so on, until the available elements can be employed directly in the construction of the desired program.

Byte: a group of eight 'bits' of digital data.

Cartography: the art and science of map and chart making.

Channels: digital data such as Band Ratios, Principal Components, multispectral bands, that are available to an image processing system.

Character map: this is a classification map or single-channel image representation, produced on a computer lineprinter with each pixel being printed as an appropriate alphanumeric symbol.

Class separability: the ability for a class to be identified or separated from all other classes.

Clustering: in multispectral feature space, classes can be identified by clusters of 'pixels'. Using a clustering classification algorithm, the computer automatically locates these clusters and uses this information to classify each pixel into a class.

Colour balance: to adjust (the intensity, hue and saturation), via the balance of the red, green and blue intensities which constitute the individual colours that make up a photographic image or video picture etc, until the required balance is achieved. A grey scale is often used to act as reference to the coloured image.

Colour infrared film: colour photographic film that is sensitive to the infrared (0.7–1.0 μm wavelength) part of the electromagnetic spectrum.

Computer compatible tape: CCT, a computer tape on which data (i.e. from Landsat etc) are stored in a suitable form so that they can be read by a computer.

Confidence level: level of confidence in the cited classification accuracy. This is usually expressed as being X% confident that at least Y% of the pixels in an area have been correctly classified. That is, for 99.9% confidence in a classification accuracy of 94.5% we would expect in only one case out of 1000 to find an area of pixels having less than 94.5% of the pixels classified correctly.

Contrast ratio: the relative difference, in grey level terms, between adjacent sampling units (or pixels) for different types of ground cover. Contrast ratio is a function of the spectral region being studied and is usually expressed in percentage units of relative change between adjacent targets.

Control: GCP, Ground Control Points; features on the ground which have been measured in a coordinate system and can also be identified on the imagery.

Convolution: the process of multiplying the values of some interpolation function by another spatial function (here considered as adjacent pixel values) to give a revised value for the location being considered. It is a technique used in resampling an image.

Coordinates: quantities (magnitudes) used to fix the position of a point in a given reference or grid system.

Covariance: a measure of the association between two variables.

Cursor: an aiming device. Commonly it is a light spot or crosshairs on a video monitor used for pointing to the location of points.

Databank: this is regarded here as a repository for data, not necessarily digital, but usually pertaining to a particular topic.

Database: this is taken as an integrated collection of datasets. These datasets, or files, may be integrated temporarily for the scheduled analysis or held together more permanently.

Dataholding: this is more at a group or agency level and describes the total aggregation of data from different sources and of different types etc. As such it would be built up from more specific subject-oriented databanks.

Dataplane: a two-dimensional representation (raster, vector or series of points) for a single factor or channel of data. As such it is a tighter definition than that for a dataset which may include more than one dataplane.

Dataset: a file of digital data.

Destriping: the process of adjusting pixel values to eliminate striping across an image. Striping occurs because the individual sensors have different gain and performance characteristics. For Landsat 1, 2 MSS, each scan is made up of six sensors hence the striping can occur in groups of six lines across the image.

Digital converter: converts analogue video signals to digital data.

Digital imagery: every sampled element in an image is given a numerical value which represents its grey tone/intensity (see figure 6.1).

Digitisation: the digitisation of a smoothly varying grey scale target by a sensor's electronic package results in the allocation of discrete numbers to specific sections of the grey scale. These numbers may be represented as a number of 'bits' for further processing. The capacity of the computer or tape or telemetry storage system sets the number of bits that can be handled per pixel and, as a result, the number of levels into which the original signal may be divided. This level of bit allocation, or digitisation, can vary between sensor systems.

Divergence: an analysis concept that quantifies class separability.

Dynamic range: a ratio of the maximum measurable signal to the minimum detectable signal (of the sensor system).

Edge enhancement: a computer process to emphasise changes in tone at boundaries between radiance regions in an image.

Environment: an external condition, or interaction of conditions, within which a system operates. These conditions are usually specified by value or quantity or range of values or quantities.

ERTS: Earth Resources Technology Satellite, the name first given to the Landsat satellite which was put in space by the United States National Aeronautics and Space Administration (NASA). It recorded images of the earth using a multispectral scanner.

Feature space: any multichannel set of measurements of a pixel's intensity in different wavelengths can give a set of *n* coordinates, for *n* measurement channels. These *n* values define a point in *n*-dimensional space—commonly referred to as *n*-dimensional feature space.

Field of View: the smallest solid angle through which the instrument is sensitive to radiation. For a scanning instrument this angle is usually measured in milliradians while for a camera it would be in degrees (see *Instantaneous Field of View*).

Filter, digital: an algorithm for removing unwanted values from digital data.

Filter, optical: a material which by absorption or reflection prevents unwanted radiation from being transmitted through the optical system.

Gain: the increase in signal power obtained during transmission from one point to another. Usually measured in decibels.

Gaussian assumption: see *Normal assumption.*

Geographic Information System: GIS, this term is taken as meaning the amalgam of integrated and/or cross referenced datasets and the necessary software to manipulate and analyse the database into products. The data may be raster or vector or points. They may be static, quasistatic or time varying. The important point is that they are registered to some unifying grid reference system. An important distinction is also made between the *actual data* in a GIS and a *dictionary* to the various types of datasets and databases within that GIS.

GCP: Ground Control Point, see *Control*.

Graticule: in a remote sensing application, this is a network of longitude (meridians) and latitude (parallels) lines which comprise the system of geographical coordinates upon which a map is drawn.

Grid: a uniform pattern of rectangular lines superimposed on maps, mosaics, or photographs for defining the coordinate positioning of points.

Grey scale: a calibrated sequence of grey tones ranging from black to white.

Ground information: data gathered from field visits or existing records and used to assist in interpreting imagery or introducing ground species coding into the computer classification process for producing thematic maps.

Ground resolution: the smallest area on the ground that can be identified in the image.

Ground truth: see *Ground information*.

Histogram: the graphical display of a set of data which shows the frequency of occurrence (along the vertical axis) of individual measurements or values (along the horizontal axis); a frequency distribution.

Homogeneity (Homogeneous): ground cover of the same kind or same mixture.

Hue: that attribute of a colour by virtue of which it differs from grey of the same brilliance and which allows it to be classed as red, yellow, green, blue or shades of these colours.

Hydrographic information: that information on a map that relates to water, i.e. streams, rivers, lakes etc.

Image: the presentation of a scene by optical, electro-optical, opto-mechanical or electronic means. The term is generally used when the scene has been recorded directly onto magnetic tape (i.e. video, digital) as against directly onto photographic film.

Image analysis: the process of analysing the data contained in an image and extracting information about the land/sea cover it represents and drawing conclusions from this extracted information.

Image enhancement: a process (digital or photographic) for highlighting the required information in an image to facilitate better photo interpretation of the image.

Imagery: the products of image-forming instruments, analogous to photography.

Infrared: that part of the electromagnetic spectrum between 0.7 and 20 μm wavelengths. This region is sometimes further subdivided into near infrared (0.7–1.3 μm), middle infrared (1.3–3.0 μm) and far infrared (7.0–15.0 μm).

Instantaneous Field of View: IFOV, the smallest solid angle through which a detector is sensitive to radiation. In a scanning system this refers to the angle subtended by the sensing system when the scanning motion is stopped. The IFOV is commonly expressed in milliradians (see also Field of View.)

Integrated Geographic Information System: an IGIS is taken as one that has a greater emphasis on dynamically integrating and interrelating various databases than may be usually included in a GIS. The distinction is regarded as somewhat flexible.

Interactive image processing: the method of data/image processing in which the user gives commands to a computer via a terminal, receives information/images displayed back in a short time and generally 'converses' with the computer.

Interclass separation: the ability for a class to be identified or separated from another class. This can also refer to the measure of separation.

Intrinsic dimensionality: the minimum number of bands/variables that are required to represent the majority of the information in an image. These bands may be chosen from the original sensor bands and/or from bands that have been created by such transformations as Principal Component, Band Ratio etc.

Irradiance: the measure in power units (W cm^{-2}) of radiant flux incident upon an object or surface.

Knowledge Based Systems: KBS, a vital component of AI inasmuch that the computing system operates on data fed to it according to some previously inserted rules or knowledge. This knowledge, like the medical diagnostic systems, is usually input in a set format and in a hierarchical manner. To operate on the data, which are usually numerical, quantified limits (or gates) must be set—and so the knowledge base is built up.

Land: this consists of the integrated components of soil, hydrology, climate, topography, geology and vegetation which together form the basis for human use of the environment. As such the term 'land' is taken as covering land cover, the sub-strata (soils, geology etc) and the use of that land (farming, forestry, grazing etc).

Land cover: the surface cover of the earth. It may be bare, vegetated, water covered, composed of buildings etc. Land cover is usually the most amenable to monitoring by remote sensing techniques.

Land information: land cover and land use data are combined with other information appropriate to the nominated region of the land surface. This extra information may include: economic/subsidy/marketing/farm practice/meteorological patterns/geological data etc.

Land use: the surface of the land can be used in various ways. Here the land cover data are combined with knowledge or other map data on land cover management practice and the appropriate land use derived.

Landsat: a satellite series launched by the United States National Aeronautics and Space Administration (NASA) to record images of the earth using a multispectral scanner (see also *ERTS*).

Likelihood: in the classification of remotely sensed images the likelihood is the probability that a classified pixel belongs to a probability distribution typified by certain parameters, considered as a function of the parameters rather than of the observation. The Maximum Likelihood classification method estimates parameters in statistical models by maximising the likelihood of observing the data with respect to the parameters of the model. The values taken by the parameters at the maximum are known as Maximum Likelihood estimates.

Lithographic process: a printing process that uses the properties of grease and water to transfer ink from an engraved surface (usually aluminium) of the image to be printed, onto paper.

Macroscopic: in the classification sense an overview classification. Each class is then regarded as an amalgam of subclasses.

Majority decision: in the classification of imagery each pixel is classified into a class. A defined area, such as a field, is allocated to the class that has the greatest number of like pixels within its boundary.

Mapping polynomial: an algorithm that defines the transformation of a map from one coordinate system to another system.

Matrix: a rectangular table containing X rows and Y columns to display an array of $X*Y$ numbers. Each term in the array can be referred to by position; for example, the element A_{12} is the term in row 1, column 2 of matrix A.

Maximum Likelihood rule: a statistical decision criterion, based on Gaussian statistics, to allocate each pixel into its most likely class (see also *Likelihood*).

Microscopic: in the imaging sense, the smallest detail that can be seen in the imagery.

Model: a model is defined as a representation of the real situation. As such it imitates, as far as possible, the actual structure or function of the natural world or mechanism. Any model has the following components: data, analysis, action/reaction mechanisms and representation systems. As commonly thought internationally, a model of a land cover or economic system represents that real life system by a set of interactive mathematical processes within a computer. These processes are fed by data streams and respond to variable conditions imposed by analysts.

MSS: see *Multispectral scanner*.

Multichannel feature space: feature space having a number of dimensions, usually equal to the number of channels being studied (see also *Feature space*).

Multimodal distribution: a histogram of classification statistics that has more than one peak in the classification application. This usually results from combining more than one class or type of measurement within the class to be studied.

Multispectral: two or more spectral bands.

Multispectral (line) scanner: an instrument that uses an oscillating or rotating mirror to scan across a scene while being moved forward by the carrying platform. It simultaneously measures, on an array of detectors, the amount of radiation in several wavelength bands.

Multitemporal: (i) scenes; two or more images that have been recorded over the same area but at different times/dates. (ii) analyses; studies using multitemporal scenes.

Multivariate analysis: a data analysis approach that makes use of multidimensional interrelationships and correlations within the data and between the data variables for effective discriminations.

Nadir: the point on the ground vertically below the remote sensing system, in a direct line between the system and the earth's centre of gravity.

Nearest neighbour: in image resampling (used in enhancement, registration etc) the multispectral values for the new pixel are obtained from its closest neighbour.

Non-parametric: An algorithm which uses no parameters to approximate the 'real world' situation. The algorithm operates on all the data presented to it.

Normal assumption–Gaussian assumption: this assumes that a histogram for any class can be approximated by a Gaussian (or Normal) probability density function. Using this assumption, classes can be described by the mean vector for the class over the channels, and the covariance matrix between channels for the class.

Orbit: the path of a satellite around the earth.

Panchromatic film: one that is sensitive to a wide spectral band, such as all visible light.

Parallelepiped classifier: a classification algorithm that partitions multidimensional feature space by defining a class region in three dimensions as a rectangular box, or a combination of boxes. If greater than three dimensions the 'box' analogy is upgraded.

Parameter: a value that remains constant during the processing of an algorithm but which may be varied each time the algorithm is processed.

Parametric: a representation, or model, of the 'real world' situation. For simple representations parameters may be estimated from the statistics of the sample, such as the mean and variance.

Pitch: rotation of an aircraft which causes a nose-up/nose-down attitude.

Point data: usually refer to a single numerical value for a given or specified point in two- or three-dimensional space. Such values can be time varying.

Principal Components: a transformation of the multispectral feature space to the position where maximum variations in the data are observed. Usually a decrease in the intrinsic dimensionality of the data is observed.

Projection (map): the representation on a plane surface (map) of the whole of, or part of, the earth's surface.

Radiance: the radiant power of an electromagnetic energy source scattered or emitted from a surface after irradiance by the source. Measured in $W \, cm^{-2} sr^{-1}$.

Radiance fall-off: the reduction in radiance caused by atmospheric absorption and reflection.

Radiance range: the range of radiance units recorded via a detector system, selected for further analysis appropriate to one particular target.

Raster: the scanning pattern used to write an image, line by adjacent line, on a cathode ray tube or on photographic film.

Raster data: numerical data that are laid down line-by-line to form a two-dimensional display are known as raster data. Along each line there is usually a constant number of picture elements (or pixels) each having specific individual numerical values.

RBV: Return Beam Vidicon, a panchromatic television system that flew on the early Landsats.

Rectification: the process of removing the effects of tilt, relief and other known distortions from imagery or photography.

Reflectance: (i) a measure of the ability of a surface to reflect energy; (ii) the ratio of the radiant energy reflected by a body to that incident upon it.

Registration: the process of superimposing two or more images so that geographic points coincide. The images may be multitemporal MSS images, map and MSS image, photographic etc and the process can be done photographically or digitally. The digital process usually involves a mapping polynomial which distorts one of the images to fit the other. If done interactively, rather than using an explicit single mapping polynomial, the technique may be known as the 'rubber sheet' method.

Resampling: the process whereby the original data form the base from which a new image is created. This new image may be at a larger or smaller scale than the original image, in which case divisive interpolation or averaging, respectively, may be used. Also the base image may be mapped onto a more geometrically correct format or onto another projection. The reallocation of radiance values to the output pixels from the input image is usually effected by a resampling process.

Resolution: the smallest spatial area on the ground that can be identified on the imagery. If spectral resolution is involved it refers to the smallest radiance range over that spectral band that can be discriminated. This may then be related to differentiating ground cover types.

Roll: rotation of an aircraft to cause a wing-up or wing-down attitude.

Sample: a sub-set of a population selected to obtain information concerning the characteristics of the population.

S-bend distortion: a distortion found in aircraft scanner imagery, which results in features on the ground lying perpendicular to the flight being distorted into an 'S' format.

Scan line: the part of an image recorded by a single detector during one sweep by the scanner.

Scanner: see *Multispectral (line) scanner.*

Scene: the area on the ground that is covered by an image or photograph.

Signatures: any characteristic, or series of characteristics, by which an object can be recognised.

Spatial information: information that is conveyed by the location of, proximity to, or orientation of, objects with respect to one another.

Spatial resolution: the smallest object on the ground that can be seen on the image. It is an explicit function of the sensor's IFOV and an implicit function of the contrast ratio, in that spectral band, of the range of ground targets presented to the system (see also *Resolution*).

Spectral information: information that is conveyed by the spectral response of individual resolution cells in the scene.

Spectral resolution: the smallest amount of spectral change that can be detected by the sensor. It is a function of both the location and breadth of the wavelength interval being accepted by the sensor's detectors, plus the level of digitisation of the incoming signal performed by the sensor (or number of distinct grey scale intervals possible in a photographic product) (see also *Resolution*).

Spectral space: see *Feature space*.

Spectrum: radiation energy arranged in continuous sequence according to wavelength or frequency.

spot: Système Probatoire d'Observation de la Terre, the French satellite system carrying the hrv (High Resolution Visible) instrument.

Standard deviation: in Gaussian statistics the standard deviation is a measure, in each individual channel of multichannel space, of the spread in the statistics for the class, or target area, being studied in that space.

Stereo photographs: two overlapping photographs that may be viewed stereoscopically (see *Aerial photography*).

Structured programming: permits the matching of individual tasks more readily to appropriate sectors of databases and machines. When applied to the actual coding itself, the concept breaks the problem into component sections, which then make more effective use of system capabilities and resources than either in-line coding or the extensive use of subroutines. The input and output modules feed into, and from, the analysis module acquiring and releasing resources for sub-set times rather than holding them for any inactive periods. Similarly the data structures are matched and progressively modified to better suit the machine segments as the process continues.

Sun-synchronous satellite: a satellite that has an orbit in which the orbital plane is near polar, such that the satellite passes over places along the same latitude at the same local sun time each day for passes in the north- or south-bound directions.

Supervised classification: a computer process that assigns each pixel to a class according to a specified decision rule, where the possible classes have been defined explicitly by the user based on training fields of known ground cover.

Synthetic channel: digital data derived from a combination of the original multispectral bands: i.e. Band Ratios, Principal Components etc.

Target: an area or object or class on the ground which is of specific interest.

Team programming: the 'team programming' concept brings together specialists in: system, input/output, database, particular analysis areas etc, to frame the package in a coordinated manner. Transportability and generality are generally both aided.

Test area: an area defined to the image processing system which is classified to check the likely accuracy of an overall classification.

Threshold: some classification systems can assign *every* pixel to a specific class no matter whether or not it is actually a member of that class. However in so assigning each pixel, each pixel has a likelihood of correct classification also attached to it. The thresholding process excludes those pixels below a nominated likelihood from the finally accepted classification product.

Top–down coding: top–down coding defines the total problem in terms of selected tasks and orders the tasks in a hierarchical order. The order itself is not of so much importance, but rather it is its place in the overall analysis path. Within each of the tasks the elements required in the task are firstly defined and this progresses to the necessary variables to support those elements and then to the language that actually defines those variables within the programme. At each stage the undefined components from a previous stage are also defined.

Topographic: the representation of natural and man-made features on a map.

Transformation: the conversion of vectors from one coordinate system to another coordinate system. The process can be linear or non-linear and involve translation and scaling. In remote sensing the term usually refers to the transformation of multispectral feature space into another feature space, or registering one image onto another image.

Training field: an area of known identity or ground type which is used in the classification process to define class boundaries within feature space. These boundaries are then used to assign each pixel in the image, sub-image or test area to its appropriate class.

Typographic (map) data: character information on a map, i.e. names, descriptions, references etc.

Unimodal distribution: a histogram that has one peak. See *Multimodal distribution.*

Unsupervised classification: see *Clustering.*

User: the consumer of the information being produced by the data processing system.

User friendly: a system that has been designed to interact more easily with the non-system-orientated user. Emphasis has been placed on the man–machine interface.

Variance: a measure of the dispersion of individual unit values about their mean. In Gaussian statistics it is the square of the standard deviation (see also *Standard deviation*).

Vector: a quantity having magnitude, sense and direction. The measure of the location of a point in space relative to another point.

Vector data: in two-dimensional space each line segment may be represented by two coordinates at each end. These four numbers define such a two-dimensional vector, of non-specific straight line length. These numbers are the vector data representation of the line segment.

Wavelength: the distance between successive wave crests, or other equivalent points, in a harmonic wave.

Weak class pairs: classes that are not well separated in feature space (see *Class separability and Divergence*).

Wide band video taperecorder: a video taperecorder capable of recording imagery at a higher density than normal taperecorders.

Yaw: rotation of an aircraft about its vertical axis causing the longitudinal axis to deviate from the flight line.

Zero-level offset: data of interest may have their lower level somewhat higher than the instrumental data zero. For some purposes a remapping of the data can occur where a new zero level is taken at the lower bound to the data region of interest.

Zero–one loss function: a function that reduces the chance of misclassifying a pixel in the Maximum Likelihood system. It assumes that a pixel is either of the class being sought or not; there is no allowance for mixtures of classes.

Bibliography

A very brief list of some texts that could support users starting in this field follows. It is obviously not meant to be exhaustive but rather to act purely as 'discussion starter' material for users.

As an introduction to general concepts:

Short N M 1982 *The Landsat Tutorial Workbook—Basics of Satellite Remote Sensing* NASA publ. No 1078 (Washington, DC: NASA)

As a further general outline of applications, sensor systems, and techniques:

Manual of Remote Sensing (2nd edn, 1983) (Falls Church, VA: American Society of Photogrammetry)

For more general, and highly readable, support to the concepts developed here:

Hord R M 1982 *Digital Image Processing of Remotely Sensed Data* (New York: Academic)

Moik J G 1980 *Digital Processing of Remotely Sensed Images* NASA publ. No SP–431 (Washington, DC: NASA)

Swain P H and Davies S M 1978 *Remote Sensing: The Quantitative Approach* (New York: McGraw-Hill)

Appendix: Sources of Material

As mentioned in the preface this book is based upon the following report.

Computer Classification of Landsat and Aircraft Scanner Images—The Collected Papers of the ERMAN Project (1982) ed. I L Thomas, Report No 766 (New Zealand: Physics and Engineering Laboratory, DSIR)

Chapter 1   New chapter
Chapter 2   Based on Chapter 5 of the above report
             Original authors: I L Thomas and S M Timmins
Chapter 3   Based on Chapter 10 of the above report
             Original authors: I L Thomas and S M Timmins
Chapter 4   Based on Chapter 2 of the above report
             Original author: N M Davis
Chapter 5   Based on Chapters 18 and 19 of the above report
             Original authors: A F Cresswell, S M Timmins and M J McDonnell
Chapter 6   Based on Chapters 5 and 6 of the above report
             Original authors: I L Thomas, V M Benning, N P Ching and S M Timmins
Chapter 7   Based on Chapter 8 of the above report
             Original authors: I L Thomas and G McK Allcock
Chapter 8   Based on Chapter 16 of the above report
             Original authors: P J Ellis, N P Ching and V M Benning
Chapter 9   Based on Chapters 3, 7 and 20 of the above report
             Original authors: I L Thomas, D W Beach and B J Winters
Chapter 10   Based on Chapter 9 of the above report
             Original authors: V M Benning, I L Thomas and N P Ching
Chapter 11   Based on Chapter 11 of the above report
             Original authors: I L Thomas and G McK Allcock
Chapter 12   Based on Chapter 13 of the above report
             Original author: N P Ching
Chapter 13   Based on Chapter 14 of the above report
             Original author: V M Benning

Chapter 14   Based on Chapter 12 of the above report
               Original authors: P J Ellis, R L Bennetts, J E Lukens and S
               M Timmins
Chapter 15   Based on Chapter 22 of the above report
               Original authors: S M Timmins, R L Bennetts and P J Ellis

Index